Lecture Notes Electrical Engineering

Volume 19

Igor V. Minin · Oleg V. Minin

Basic Principles of Fresnel Antenna Arrays

Springer

Igor V. Minin
Novosibirsk State Technical University
Dept. Information Protection
Prospekt Karla Marksa 20
Novosibirsk
Russia 630092
igor.minin@ngs.ru

Oleg V. Minin
Novosibirsk State Technical University
Dept. Information Protection
Prospekt Karla Marksa 20
Novosibirsk
Russia 630092
Prof.minin@gmail.com

ISBN: 978-3-540-79558-2 e-ISBN: 978-3-540-79559-9

Library of Congress Control Number: 2008927519

Cover design: eStudio Calamar S.L.

Printed on acid-free paper

9 8 7 6 5 4 3 2 1

springer.com

Dedicated to the father, friend and colleague, all in one, and equally to our mother. Without their help and support, this book would never have been written.

Foreword

The interest in Fresnel antennas has been growing steadily since the late 1980s. The simple but ingenious Fresnel zone concept has been extended greatly and exploited to create different types of Fresnel zone antennas including the offset Fresnel zone antennas, the phase correcting Fresnel lenses and zone plate antennas, and reflect arrays. This is primarily driven by the increasing demand on low-cost, low-profile, and high-gain antennas.

I know Professor I. Mimin and Professor O Mimin "over the net" since early 1990s, and I have always been impressed by their untiring effort to pursue various new ideas related to Fresnel zone diffraction and Fresnel zone antennas. It is encouraging to see the publication of their new book, "*Basic Principles of Fresnel Antenna Array*," which covers the research on Fresnel antenna arrays. For a given antenna aperture, employing an array of Fresnel antennas as opposed to a single one can reduce the distance between the feed and the aperture, thereby reducing the profile of the overall system. On the other hand, a feed network is required to connect all the feeds, and the losses in the feed network need to be managed. It is expected the book will attract more interest in Fresnel zone antennas and stimulate new ideas on this fascinating subject.

Dr. Y. Jay Guo
Clayton South, VIC

Foreword

This work is an important addition to the field of array antennas. It also extends the knowledge and applications of Fresnel zone plate lens antennas (FZPLAs) and is the first study of FZPLAs in arrays. Profs. Igor and Oleg Minin have done extensive research and development on FZPLAs. They are among the world's most famous researchers of the Fresnel zone plate antennas, having written four books (two in Russian) and numerous published articles. The Drs. Minin have made extensive millimeter-wave measurements on Fresnel zone plate antennas and rigorous comparisons with theory. This book is an excellent summary of the state of the art of diffractive optical elements (DOE) and introduces the application to array antennas. The book is very comprehensive and thorough, and offers innovative new ideas.

Most of their work is applicable to zone plates having comparable diameters and focal lengths, the so-called large-angle zone plate, which is the configuration most often used at microwave, millimeter-wave, and terahertz frequencies. This book addresses the need for low-cost, low-profile, and lightweight antennas. A new hexagonal FZPL antenna is described. This has the advantage over the usual circular cross-section FZPLA in that its shape fits more compactly into an array. A technique involving Fresnel rotation is described. This improves the radiation pattern of the hexagonal FZPL antenna. In addition, the concept of changing the reference phase of the Fresnel zone radii from 0° to a more optimal number up to 180° is described. These last two ideas improve the radiation characteristics.

The technology of diffractive optics in the microwave to terahertz frequency range has seen much activity in recent years. This is the fifth book to be published on FZPL antennas, and in the past two decades some 100 research articles have appeared in the worldwide literature. The field of diffractive optical elements has been used in many applications, including radar, radiometry, missile terminal guidance seekers, point-to-point communications, and field tests of atmospheric effects, especially in cases where low cost, low attenuation, low weight, low physical volume, and ease of manufacturing are the considerations.

This book begins with a review of elementary basics of antennas, defining such quantities as radiation power density and intensity, directivity, effective area of an antenna, and radiation pattern, parameters relevant to any antenna. Subjects relating to arrays are also described, including such topics as array factor, diversity combining, and grating lobe criteria. Fresnel zone plate antennas are also reviewed, comparing lenses and zone plates and giving Fresnel zone plate antenna design information. This includes the information about reflector-backed FZPLAs and reconfigurable zone plates utilizing photoconductive material or mechanical shutters. One chapter considers FZPLA candidates for arrays based on geometry:

rectangular, square, circular, polygonal, hexagonal, star-shaped, and arbitrary-shaped apertures. Perforated dielectric Fresnel lenses are considered in another chapter.

A final chapter treats arrays of small lenses, lens arrays based on Luneberg lenses, fly's-eye imaging concepts, waveguide lens array technology, and several other unusual applications. The net result is an extremely diverse coverage of numerous configurations and applications. The book should see broad use and wide application. Profs. Igor and Oleg Minin have again produced a significant document that will be valuable to antenna designers and those designing systems such as radar or communications.

Dr. James C. Wiltse
Atlanta, Georgia

Contents

Introduction

Wireless and satellite communications are vital in the daily activities of the average individual and business. Most services can now be received from anywhere at anytime. These services must deliver wireless high bandwidth multimedia signals which provide reliably simultaneous access to voice, fax, high speed internet, and video. Also homeland security, nondestructive testing, imaging technologies are actually today. Antennas represent a critical technology in any of these wireless systems.

Achieving a lowcost, lowprofile, and lightweight antenna in the microwave/ millimeter wave band (MMW) is not trivial. The core problem is generally that low aperture efficiencies result with existing lowprofile technologies. The parabolic reflector, reflectarray, and dielectric hyperbolic lens are all high gain, high efficiency antennas with desirable radiation characteristics but they are not lowprofile. Each one consists of a feed placed in front of the aperture by a certain distance which causes the antenna depth to be large even at MMW. There are also stringent tolerance issues that exist in each of their construction which make them costly to fabricate at these high frequencies. Fresnel lens antennas, including Fresnel zone plates [15], are also not very low profile due to their feed being placed at the focal point in front of the aperture in a similar manner to the other high gain antennas.

The most common low profile technology is the microstrip antenna, which is typically used in an array to achieve higher gain. At higher frequencies, such as those in the MMW, the microstrip array feed network is very lossy which seriously degrades the aperture efficiency. The slotted waveguide array is another potential low profile antenna option, but its cost is prohibitive owing to the tight fabrication tolerances required at these frequencies. The only other alternative low profile antenna is the dielectric grating but, similar to the slotted waveguide array, fabrication becomes more complex and costly in the MMW.

A potential solution to the low profile antenna problem at MMW lies in using lens technology in an array. By replacing a large diameter lens with an array of smaller lenses, the overall profile of the lens can be significantly reduced. However, there are some drawbacks to lens arrays such as the requirement for more feeds and the degradation of sidelobe performance. Very little work has been performed on using lenses in arrays though, which means that this area remains relatively unexplored.

The Fresnel zone plate lens (FZPL) antenna, in particular, is an interesting candidate for the array element since it is the lowest cost, lowest profile, and lightest weight lens antenna [15]. The aperture of the FZPL antenna has the lowest profile of the Fresnel lens family of antennas because it is simply printed on a thin substrate which does not need to be of microwave grade. This also makes the FZPL antenna inexpensive to fabricate.

It could be noted that a Fresnel zone plate works by interference and a Fresnel lens by refraction. The focal length of a Fresnel lens is not proportional to the wavelength of radiation, but for a zone plate, it is.

It is interesting also to note that in the paper on optical processing for synthetic aperture array radar, Cutrona et al. [6] point out that the synthetic aperture recorded signals are one dimensional Fresnel zone-plate lenses. In the paper [7] a method of electronically processing synthetic aperture arrays utilizing the Fresnel zone-plate lens concept was also presented.

Different from the conventional lens which uses curved surfaces to refract radiation, so-called a digital lens is made of a transparent flat plate with numerous micro zones, each zone has its location and optical characteristics precisely and in common case digitally defined. When radiation passes through these zones, the light waves *interfere* with each other due to phase differences and results in light wave manipulation. This manipulation technology can be used to deliver superior quality of images and it is widely applicable to many other industries such as radar *antenna* and *communications*.

In general, the term diffractive optical elements (DOE) refer to those that are based on the utilization of the wave nature of radiation. The DOEs in common case are based on grating composition. The grating effect is dominant and defines their function and limitations. In general, grating dispersion is much stronger than prism dispersion. Thus, frequency properties (or chromatic dispersion) strongly influences to the imaging properties of DOEs. The relative conditional categorization of DOEs can be divided into several main subsections: diffractive lenses (elements that perform functions similar to conventional refractive lenses, e.g. they form images); diffractive antennas (quasi-optical elements that form arbitrary beampattern in far field); kinoforms (DOE whose phase modulation is introduced by a surface relief structure); binary optics; diffractive phase elements (DOEs that introduce phase change). The DOE can be seen either also as an aperture synthesis array. Beams from the individual apertures are recombined by diffraction and interference. The apertures (void circular, rectangles, hexagonal or more complex shapes) are positioned so that at the first order of diffraction (2π phase shift from one aperture subset to the next), an incoming plane wave is turned into a spherical outgoing wavefront.

One of the factors that have stimulated much of the recent interest in diffractive optics at any frequency waveband has been the increased optical performance of such optical elements. This allows the fabrication of optical elements that are smaller (compared to wavelength), lighter and cheaper to fabricate, are more rugged and have superior performance that the conventional optical or/and quasioptical components they often replace. Important, the design capabilities for binary optics now available can make possible the design and manufacture to components including antennas having optical and focusing properties never before produced.

FZPL antenna is one of the simple digital lenses or DOE. Flat surfaces are two dimensional, therefore much cheaper to fabricate than three dimensional contour surfaces. It's also more precise due to digitally defined microscopic zones and simple geometric shape. With the conventional technology, it's very difficult, if not possible to produce a lens array on a single sheet of material, not to mention the cost. To further lower the fabrication cost, lenses even can be *printed* on a flat surface.

The traditional circular FZPL antenna, however, does not perform well in an array. This is because the elements cannot be placed any closer together than when their edges touch at a single point, which does not provide an adequate separation to minimize grating lobes [8]. As such, a novel hexagonal FZPL (HFZPL) antenna [9] is described in details which can be more effectively packed in an array due to its shape.

Prior to considering the HFZPL antenna in an array, two main ideas described as methods to potentially improve the radiation characteristics. The first idea is to change the reference phase of the Fresnel zone radii from the standard 0° to a more optimal value between 0 and 180 [10]. To further improve the radiation characteristics of the HFZPL antenna, a technique involving Fresnel zone rotation [9] are also described. This method might also provide interesting beamshaping options when used on the elements in an array.

The book is intended to serve engineers, researchers and student in the field of antennas, micriwave/millimeter wave/THz wave engineering and telecommunications. Also the authors believe that the book will also be of interest to designers of optical systems because, with scaling effects taken into account, the characteristics of diffractive quasioptical elements are valid for diffractive focusing elements of integrated optics.

We are greatly indebted to Prof. Jim Wight (Chancellor's University), Dr. Aldo Petosa (CRC, Canada) and specially to Dr. Sara M. Stout-Grandy, the big number of results described in the book were obtained during her 3 years Ph.D. thesis work titled "Investigation of Planar Fresnel Zone Plate Antennas" under the supervision of Dr.A.Petosa, Prof I. Minin and Prof. J. Wight, for permission to use figures and some of joint results.

We are infinitely grateful to Dr. J. C Wiltse (Georgia Institute of Technology, Atlanta, USA) and Dr. Jay Guo (CSIRO), for his willingness to write the Foreword to this book and also to Christoph Baumann, editor for English engineering books at Springer Heidelberg, Germany, for constant attention and help in our work with the book.

References

1. Minin, Minin. Diffractive Quasioptics: Research and Production Association "InformTEI", Moscow, 1992 180p. (in Russian).
2. Minin. Minin. Diffractive Quasioptics and its Applications Novosibirsk: SibAGS, 1999. 307p. (in Russian).
3. Hristo D. Hristov. Fresnel Zones in wireless links, zone plate lenses and antennas –Artech House. Boston, London, 2000.
4. Minin., Minin. Diffractive Optics of Millimeter Waves. – IOPP Publisher, London-Boston, 2004.
5. Y.J.Guo, S.K.Barton. Fresnel Zone Antennas. Kluwer Academic Publishers, 2002.
6. L. J. Cutrona, E. N Leith, L. J. Porcello, W. E. Vivian On the Application of Coherent Optical Processing Techniques to Synthetic-Aperture Radar Proc. IEEE,. 54, pp. 10261032
7. R. K. Moore, J. W. Rouse. Electronic Processing for Synthetic Aperture Array RadarProceedings Letters, 1967, pp. 233234.
8. Petosa, A. Ittipoon. Thirakoune Array of Perforated Dielectric Fresnel LensesURSI International Symposium on EM Theory, EMTS 2004, Pisa, Italy, May 2327, 2004, pp. 969971.
9. Minin, Minin. Array of Fresnel Zone Plate Lens Antennas: Circular, Hexagonal with Chiral Symmetry and Hexagonal Boundary Digest of the Joint 31st International Conference on Infrared and Millimeter Waves and 14th International Conference on Terahertz Electronics, September 1822, 2006, Shanghai, China, p. 270.
10. Minin, Minin. Three Dimensional Fresnel Antennas. – Chapter 5 in: Advances on Antennas, Reflectors and Beam Control, Research Signpost, Kerala, India, 2005, pp. 115148.

Chapter 1
The Brief Elementary Basics of Antenna Arrays

For some applications, single-element antennas are unable to meet the gain or radiation pattern requirements. Combining several single antenna elements in an array can be a possible solution.

1.1 Some Basic Antenna Parameters Definitions

1.1.1 Radiation Power Density

Radiation Power density, W, gives a measure of the average power radiated by the antenna in a particular direction and is obtained by time-averaging the Poynting vector.

$$W_r(r,\theta,\phi)=\frac{1}{2}\mathrm{Re}\left[E\times H^*\right]=\frac{1}{2\eta}\left|\bar{E}(r,\theta,\phi)\right|^2 \text{ (Watts/m}^2\text{)}, \quad (1.1)$$

where, E is the electric field intensity, H is the magnetic field intensity, and η is the intrinsic impedance (Fig. 1.1).

1.1.2 Radiation Intensity

Radiation intensity, U, in a given direction is the power radiated by the antenna per unit solid angle. It is given by the product of the radiation density and the square of the distance r.

$$U=r^2W_r \text{ (Watts/unit solid angle)} \quad (1.2)$$

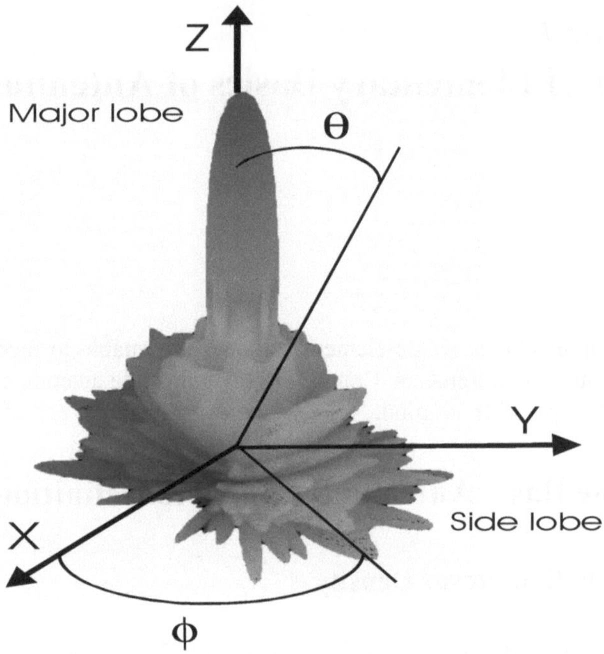

Fig. 1.1 3D polar dimensional radiation power pattern

1.1.3 Total Power Radiated

The total power radiated by the antenna in all the directions is given by:

$$P_{tot} = \int_0^{2\pi}\int_0^{\pi} W_r(r,\theta,\phi)r^2 \sin\theta d\theta d\phi \tag{1.3}$$

$$= \int_0^{2\pi}\int_0^{\pi} U(\theta,\phi)\sin\theta d\theta d\phi \quad \text{(Watts)} \tag{1.4}$$

1.1.4 Directivity

The Directive gain, D_g, is the ratio of the radiation intensity in a given direction to the radiation intensity in all the directions, i.e.:

$$D_g = \frac{4\pi U(\theta,\phi)}{P_{tot}} = \frac{4\pi r^2 W_r(r,\theta,\phi)}{\int_0^{2\pi}\int_0^{\pi} W_r(r,\theta,\phi) r^2 \sin(\theta) d\theta d\phi} = \frac{4\pi U(\theta,\phi)}{\int_0^{2\pi}\int_0^{\pi} U(\theta,\phi)\sin(\theta) d\theta d\phi} \quad (1.5)$$

The Directivity, D_0, is the maximum value of the directive gain, D_g, for a given direction, i.e.:

$$D_0 = \frac{4\pi U_{max}(\theta,\phi)}{P_{tot}}, \quad (1.6)$$

where $U_{max}(\theta,\varphi)$ is the maximum radiation intensity.

1.1.5 Effective Area of an Antenna

Consider that an EM wave of power density, P, is intercepted by an antenna of actual physical area, A. The EM wave will induce a voltage, V, across the radiation resistance of the antenna. The antenna can deliver a maximum power, W_{max}, to a load, $R_L = R_{rad}$, i.e., a load matched to the antenna radiation resistance (assuming ohmic losses are zero). The maximum power that the antenna can deliver to a load is:

$$W_{max} = \frac{V^2}{2R_{rad}}. \quad (1.7)$$

The effective area, A_e, of an antenna is defined as the area required for gathering enough power from the incident power density, P, of the EM wave in order to deliver power, W_{max}, to the load.

$$A_e = \frac{W_{max}}{P}. \quad (1.8)$$

Let us find the effective area of a short dipole. The radiation resistance is $R_{rad} = 80\pi^2\left(\frac{l}{\lambda}\right)^2$.

The power density at the short dipole is $P = \frac{E^2}{\eta_0} = \frac{E^2}{120\pi}$.

From (1.8) $A_e = \frac{V^2}{4PR_{rad}} = \frac{V^2(120\pi)}{4E^2(80\pi^2)}\left(\frac{\lambda}{l}\right)^2 = \frac{3}{8\pi}\left(\frac{V\lambda}{El}\right)^2$.

Now, the voltage induced across a short dipole is just $V = El$. So the effective area is $A_e = \frac{3}{8\pi}\lambda^2$. The directivity of a short dipole is $D_{max} = 1.5$. Then the relative directivity of a short dipole is:

$$\frac{D_{max}}{A_e} = \frac{1.5}{\left(\frac{3\lambda^2}{8\pi}\right)} = \frac{4\pi}{\lambda^2}.$$

1.1.6 Radiation Pattern

The Radiation Pattern of an antenna can be defined as the variation in field intensity as a function of position or angle. Let us consider an anisotropic radiator, which has stronger radiation in one direction than in another. The radiation pattern of an anisotropic radiator consists of several lobes. One of the lobes has the strongest radiation intensity compared to the other lobes. It is referred to as the Major lobe. All the other lobes with weaker intensity are called Minor Lobes. The width of the main beam is quantified by the Half Power Beamwidth (HPBW), which is the angular separation of the beam between half-power points.

1.2 Antenna Arrays, Radiation Pattern, and Array Factor

Let us consider briefly the basics of antenna arrays following [1, 2]. The antenna elements can be arranged to form a 1, 2 or 3 dimensional antenna array. As a rule in practice either a 1 or 2 dimensional antenna array is used.

The overall radiation pattern changes when several antenna elements are combined in an array. An array consists of two or more antenna elements that are spatially arranged and electrically interconnected to produce a directional radiation pattern. The interconnection between elements, called the feed network, can provide fixed phase to each element or can form a phased array. The geometry of an array and the patterns, orientations, and polarizations of the elements influence the performance of the array. The so-called ‘array factor’ is used: this factor quantifies the effect of combining radiating elements in an array without the element-specific radiation pattern taken into account. The overall radiation pattern of an array is determined by this array factor combined with the radiation pattern of the antenna element. The overall radiation pattern results in a certain directivity and thus gain is linked through the efficiency with the directivity. Directivity and gain are equal if the efficiency is 100%.

1.2.1 Broadside and Fire Arrays

Arrays can be designed to radiate in either broadside, i.e., radiation perpendicular to array orientation (the z-axis) or end fire, i.e., radiation in the same direction as the array orientation (the y-axis). To simplify the problem, we will focus on broadside arrays and only radiation in the z direction is considered. This allows for easy transformation to 2 dimensional planar arrays with the elements in the (xy) plane.

1.2.2 Defining Array Factor

The array factor depends on the number of elements, the element spacing, amplitude and phase of the applied signal to each element. The number of elements and the element spacing determine the surface area of the overall radiating structure. This surface area is called aperture. A larger aperture results in a higher gain. The aperture efficiency quantifies how efficiently the aperture is used. The influence of these parameters will be further explained with the aid of a linear array of isotropic radiating elements. An isotropic radiating element radiates an equal amount of power in all directions, i.e., it has a directivity of 1 (0 dB) and a gain of 1 (0 dB) if the efficiency were 100%.

A three dimensional array with an arbitrary geometry is shown in Fig. 1.2. To simplify throughout this discussion, it is assumed that the source of the wave is in the far field of the array, and the incident wave can be treated as a plane wave. In spherical coordinates, the vector from the origin to the *n*-th element of the array is given by $\bar{r}_m = (\rho_m, \theta, \phi_m)$, and $-\hat{k} = (1, \theta, \phi)$ is the vector in the direction of the source of an incident wave. Throughout this discussion, it is assumed that the source of the wave is in the far field of the array and the incident wave can be treated as a plane wave. To find the array factor, it is necessary to find the relative phase of the received plane wave at each element. The phase is referred to as the phase of the plane wave at the origin. Thus, the phase of the received plane wave at the *n*-th element is the phase constant $\beta = \frac{2\pi}{\lambda}$ multiplied by the projection of the element position $\bar{r}_m$ on to the plane wave arrival vector $-\hat{k}$. This is given by $-\vec{k} \bullet \vec{r}_m$ with the dot product taken in rectangular coordinates.

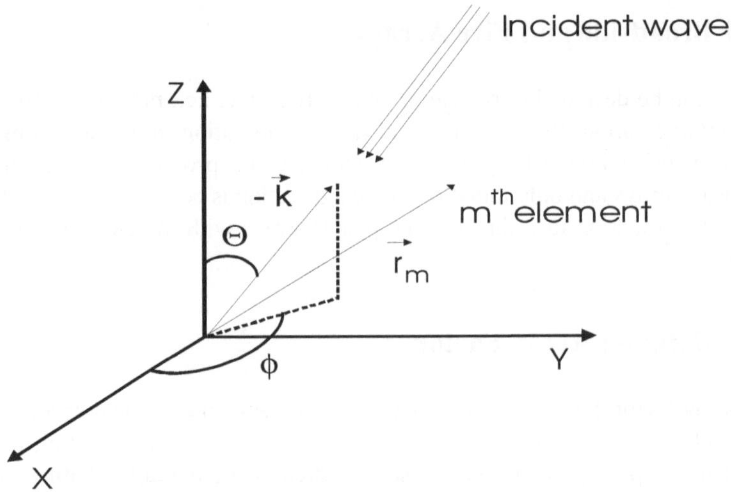

Fig. 1.2 A three dimensional array with an arbitrary geometry

In rectangular coordinates, $-\hat{k} = \sin\theta\cos\phi\hat{x} + \sin\theta\sin\phi\hat{y} + \cos\theta\hat{z} = r^2$ and $\vec{r}_m = \rho_m \sin\theta_m \cos\phi_m\hat{x} + \rho_m \sin\theta_m \sin\phi_m\hat{y} + \rho_m \cos\theta_m\hat{z}$, and the relative phase of the incident wave at the *n*-th element is:

$$\zeta_m = -\vec{k} \bullet r_m$$

$$= \beta\rho_m(\sin\theta\cos\phi\sin\theta_m\cos\phi_m + \sin\theta\sin\phi\sin\theta_m\sin\phi_m + \cos\theta\cos\theta_m)$$

$$= \beta(x_m\sin\theta\cos\phi + y_m\sin\theta\sin\phi + z_m\cos\theta).$$

For an array of M elements, the array factor is given by [3].

$$AF(\theta,\phi) = \sum_{m=1}^{M} I_m e^{j(\zeta_m+\delta_m)},$$

where I_m is the magnitude and δ_m is the phase of the weighting of the *m*-th element. The normalized array factor is given by:

$$f(\theta,\phi) = \frac{AF(\theta,\phi)}{\max\{|AF(\theta,\phi)|\}}.$$

Note this would be the same as the array pattern if the array consisted of ideal isotropic elements.

As an example, let us consider an array of any 2 identical antennas. The power pattern for the array will be the product $P(r,\theta,\phi)=F_{unit}(r,\theta)F_{array}(\psi)$, where $\psi=\beta d\cos\phi+\alpha$. For an array of 2 elements:

$$F_{array}(\psi)=4\cos^2\left(\frac{\psi}{2}\right).$$

The unit pattern is also called the element pattern. The unit pattern is usually evaluated at $\theta=\frac{\pi}{2}$. Examples are:

Hertzian dipole:

$$F_{unit}(r,\theta)=\frac{\eta_0\beta^2 I_0^2 l^2}{32\pi^2 r^2}.$$

Long Dipole:

$$F_{unit}(r,\theta)=\frac{15I_0^2}{\pi r^2}\left[\frac{\cos\left(\frac{\beta L}{2}\cos\theta\right)-\cos\left(\frac{\beta L}{2}\right)}{\sin\theta}\right]^2.$$

Consider an array of 2 half dipoles along the x-axis, with spacing $\lambda/4$ and phase delay $\pi/2$ for the drive current to the second antenna.

$$\psi=\beta d\cos\phi+\alpha,$$

$$\psi=\left(\frac{2\pi}{\lambda}\right)\left(\frac{\lambda}{4}\right)\cos\phi-\frac{\pi}{2}=\frac{\pi}{2}\cos\phi-\frac{\pi}{2}=\frac{\pi}{2}(\cos\phi-1),$$

$$F_{array}(\psi)=4\cos^2\left(\frac{\psi}{2}\right)=4\cos^2\left(\frac{\pi}{4}(\cos\phi-1)\right).$$

The normalized array pattern is $P_n(\phi)=\cos^2\left(\frac{\pi}{4}(\cos\phi-1)\right)$.

The maximum occurs for $\left(\frac{\pi}{4}(\cos\phi-1)\right)=0\Rightarrow\cos\phi=0\Rightarrow\phi=0.$

Zeros occur when

$$\left(\frac{\pi}{4}(\cos\phi-1)\right)=\pm\frac{\pi}{2}\Rightarrow\cos\phi=\pm2\Rightarrow\cos\phi=-1\Rightarrow\phi=\pi.$$

Consider an array of 2 half dipoles along the x-axis, with spacing $\lambda/2$ and phase delay $\pi/2$ for the drive current to the second antenna.

$$\psi = \beta d \cos\phi + \alpha,$$

$$\psi = \left(\frac{2\pi}{\lambda}\right)\left(\frac{\lambda}{2}\right)\cos\phi - 0 = \pi\cos\phi,$$

$$F_{array}(\psi) = 4\cos^2\left(\frac{\psi}{2}\right) = 4\cos^2\left(\frac{\pi}{2}\cos\phi\right).$$

The normalized array pattern is:

$$P_n(\phi) = 4\cos^2\left(\frac{\pi}{2}\cos\phi\right).$$

The maximum occurs for

$$\left(\frac{\pi}{2}\cos\phi\right) = 0 \Rightarrow \cos\phi = 0 \Rightarrow \phi = \pm\frac{\pi}{2}.$$

Zeros occur when

$$\left(\frac{\pi}{2}\cos\phi\right) = \pm\frac{\pi}{2} \Rightarrow \cos\phi = \pm 1 \Rightarrow \phi = 0, \pi.$$

1.2.3 Array Pattern

Importantly, if each element has a pattern $g_m(\theta,\phi)$, which *may be different* for each element, the normalized array pattern is given by:

$$F(\theta,\phi) = \frac{\sum_{m=1}^{M} I_m g_m(\theta,\phi) e^{j(\zeta_m+\delta_m)}}{\max\left\{\left|\sum_{m=1}^{M} I_m g_m(\theta,\phi) e^{j(\zeta_m+\delta_m)}\right|\right\}}. \tag{1.9}$$

In (1.9), the element patterns must be represented such that the pattern maxima are equal to the element gains relative to a common reference.

1.2.4 Effect of Array Geometry and Element Patterns

Let us consider briefly some main antenna array parameters. The array directivity increases with the number of elements. Figure 1.3 shows the directivity of 3 arrays with 2 (solid), 5 (dashed) and 10 (dotted) elements. The element spacing is 0.4 times the wavelength for all the arrays in Fig. 1.3. The presence of side lobes next to the main lobes is important: this is typical for arrays. The number of side lobes and the side lobe level increase with the number of elements. It can be noted that according to the array factor definition, there are 2 main lobes. There is a main lobe at theta 0° (positive z axis) and a main lobe at theta 180°/180° (negative z axis).

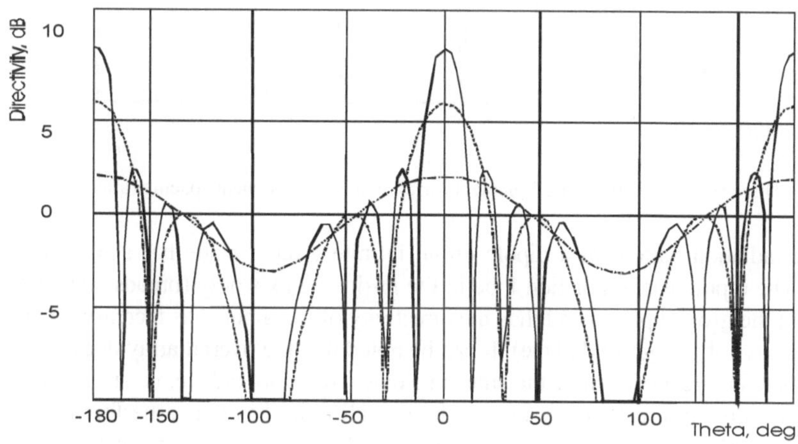

Fig. 1.3 Directivity of a 2 (*solid*), 5 (*dashed*) and 10 (*dotted*) element array with 0.4 λ element spacing (after [4])

The element spacing has a large influence on the array factor as well. A larger element spacing results is a higher directivity. However, the element spacing is generally kept smaller than $\lambda/2$ to avoid the occurrence of grating lobes. A grating lobe is another unwanted peak value in the radiation pattern of the array.

The "radiation pattern" of an isotropic element is a horizontal line. In this case, the overall radiation pattern is the same as the array factor since an isotropic element radiates the same amount of power in all directions.

Figure 1.4 shows the radiation pattern of a dipole (dotted), the same array factor (dashed) without dipoles, and the overall radiation pattern of the array with dipoles (solid). The overall radiation pattern is clearly different from the array factor, i.e., the directivity has increased with the dipole's directivity and the overall radiation pattern is slightly modified due to the dipole's radiation pattern.

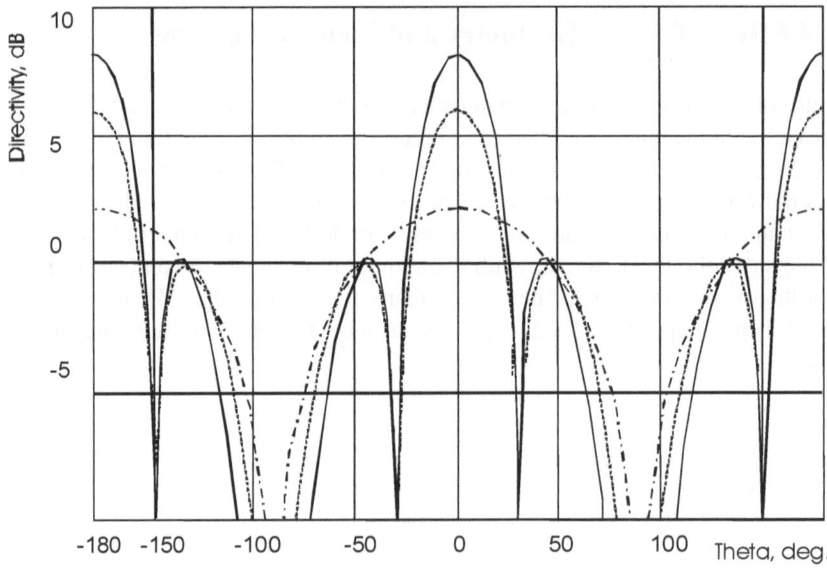

Fig. 1.4 Directivity of a dipole in a 5 elements array with 0.4λ element spacing (after [1])

The radiation pattern of a dipole on an infinite ground plane has some advantages. The dipole has a radiation lobe in the positive z axis only (broadside and because of the ground plane) and thus the directivity of the dipole has increased with 3 dB (because of that ground plane). It can be noted that the overall array does indeed not radiate in directions where the antenna element does not radiate, i.e., no radiation in the negative z direction any more. The overall array has thus a perfect front to back ratio; this makes sense because we have used an infinite ground plane.

1.2.5 Half Power Beamwidth (HPBW) (for the major lobe)

In order to compute the *HPBW* in addition to the angle of first maximum, the half-power point is also required. The half-power point can be calculated by:

$$\frac{N}{2}\psi = \frac{N}{2}(kd\cos\theta_h + \beta) = \pm 1.391, \tag{1.10}$$

$$\theta_h = \cos^{-1}\left[\frac{\lambda}{2\pi d}\left(-\beta \pm \frac{2.782}{N}\right)\right]. \tag{1.11}$$

Now the *HPBW* can be calculated as:

$$HPBW = 2\left|\theta_m - \theta_h\right|. \tag{1.12}$$

Note: The *HPBW* equation provided here is only for nonsteered arrays.

1.2.6. Feeding of an Array

In all the previously discussed arrays, the element spacing has been kept constant and the elements were fed with the same amplitude and phase. The resulting arrays were linear arrays with uniform spacing, uniform amplitude, and equal phase.

However, the power does not necessarily have to be distributed with equal amplitude and/or phase. Unequal power and phase distributions to the individual elements can be used to modify the side lobe level, directivity, and direction of the main lobe. A range of standard amplitude and phase distributions exists (e.g. uniform, cosine pedestal), but this is beyond the scope of this article. When the power distribution is optimized to reduce the side lobe level, the efficiency of the array decreases when the phase distribution is optimized to perform beam steering; new side lobes will show up as the main beam is deflected sideways.

Most literature on array antennas considers linear arrays of omnidirectional elements. Other geometries and element patterns are possible, and these factors influence array performance. It is well known that using elements with directional patterns can introduce grating nulls in patterns. Computer modeling was used in [5] to compare triangular, square, and cylindrical array configurations using directional elements. "Spatial reference" algorithms that compute the direction of arrival of desired and interfering signals, which then form directional beams were used.

1.2.7 Diversity Combining [2, 5, 6]

In addition to phased and adaptive arrays, signals from multiple antennas can be combined to improve performance in fading channels. From a collection of M antennas, the branch with the largest signal to noise ratio (SNR) at any time is selected and connected to the receiver. As one would expect, the larger the value of M, the higher is the probability of having a larger signal to noise ratio at the output. Maximal ratio combining takes a better advantage of all the diversity branches in the system. The branches are then co-phased prior to summing in order to ensure that all branches are added in phase for maximum diversity gain. The summed signals are then used as the received signal. Maximal ratio has advantages over selection diversity but is more complicated; proper care has to be taken in order to ensure that signals are co-phased correctly and gain coefficients have to be constantly updated. A variation of maximal ratio combining is equal gain combining. In this scheme, the gains of the branches are all set to the same value and are not changed thereafter. Also the output is a co-phased sum of all the branches.

1.2.8. Array Response Vector

The array response vector for a signal with direction of arrival (θ, ϕ) and polarization state, P, can be written as follows:

$$a(\theta,\phi,P)=\begin{bmatrix} e^{j\zeta_1}g_1(\theta,\phi,P) \\ e^{j\zeta_2}g_2(\theta,\phi,P) \\ . \\ . \\ . \\ e^{j\zeta_m}g_m(\theta,\phi,P) \end{bmatrix}.$$

The phase shifts, ζ_m, represent the spatial phase delay of an incoming plane wave arriving from angle (θ,ϕ). The factor $g_m(\theta,\phi,P)$ is the antenna pattern of the m-th element.

1.2.9 Spatial-Polarization Signature

The spatial-polarization signature is the total response of the array to a signal with N multipath components and is expressed as:

$$v=\sum_{n=1}^{N}\alpha_n a(\theta_{n,}\phi_n,P_n),$$

where α_n is the amplitude and phase of the n-th component. The angle of arrival and polarization state of the n-th component are given by θ_n,φ_n, and P_n.

1.2.10 Grating Lobe Criteria

The textbook grating lobe criteria [7, 8] that restricts the scan angle of a phased array in order to eliminate grating lobes from entering "real space" is defined as:

$$\sin\theta \le (\lambda/d)-1,$$

where θ is the scan angle of the array, λ the operating wavelength, and d the spacing between elements. Such a condition eliminates any beam, other than the main beam, from forming in visible space and the antenna designer need not be concerned about radiating energy or receiving energy (other than normal sidelobes) from unwanted directions. This is a desirable condition (no grating lobes in real space), but imposes a strict limit on element spacing. For a non-scanned antenna – main beam on boresight – the theoretical element spacing can approach λ. For large scan - approaching 90 degrees- the element spacing: approaches a $\lambda/2$ limit. So the condition on the spacing between elements is:

$$d \le \frac{\lambda}{\sin\theta+1}.$$

1.3 Brief Review of FZP Antennas

1.3.1 Why Lens Antenna?

At millimeter wave frequencies, parabolic reflectors are typically used to realize high gains. An attractive alternative is the dielectric lens antennas, similar to the optical lens, which do not require as stringent tolerances in fabrication as reflectors do, and can be made using inexpensive plastic material. They can be used to focus the beam of lower gain feed antennas to produce highly directive patterns with low side lobe levels. Artificial dielectric material can also be used to enhance the performance of these lens antennas.

From the Fig. 1.5 it follows that a deformation of antenna surface, δ, produces a significantly smaller focus-to-aperture-plane path length error in the lens (Fig. 1.5a) than in the reflector (Fig. 1.5b).

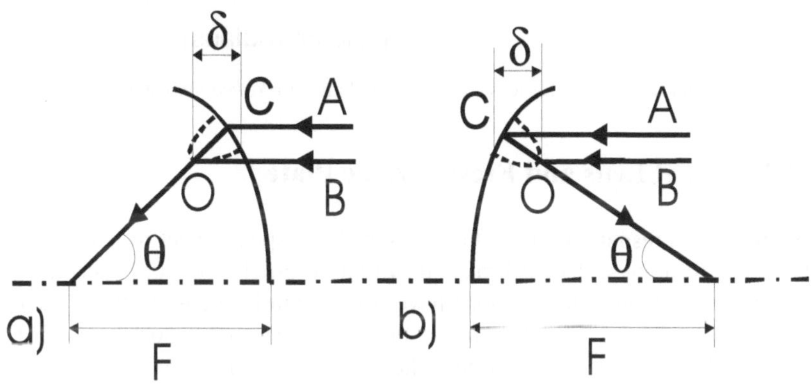

Fig. 1.5 Deformation characteristics of lens (**a**) and mirror (**b**) antennas

From the simple geometrical consideration for the path length error, we may to write that for the lens it is:

$$|ACF| - |BOF| = \delta\left(\frac{1}{\cos\theta} - 1\right)$$

and for the mirror antenna it is:

$$|ACF| - |BOF| = \delta\left(\frac{1}{\cos\theta} + 1\right).$$

In Fig. 1.6 below, the dependences of the path length error both for lens and mirror antennas vs. θ-angle (in radians) are shown. We may conclude that the lens antenna is more tolerant to surface errors than a mirror antenna. The reduced surface accuracy property of lens antenna appears to have remained relatively unappreciated.

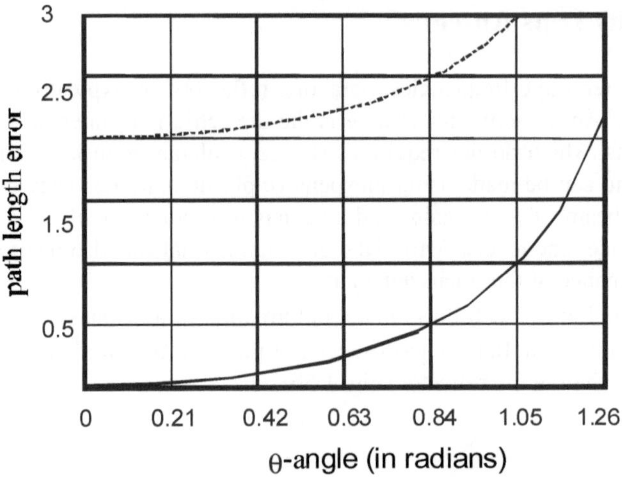

Fig. 1.6 The path length error for lens (*solid line*) and mirror (*dashed line*) antennas vs θ-angle

1.3.2 Fresnel Lens and Fresnel Zone Plate

The classical Fresnel lens is different from the shaped dielectric lens in that the phase transforming surface is broken down into Fresnel zones, or concentric circular sections. Although the sections have discontinuities between them, each section maintains the surface curvature of the original shaped dielectric lens but with significantly less dielectric material. The smaller lens profile is made possible by removing material such that the difference between ray paths on each side of the step is a full wavelength.

It could be noted that a Fresnel zone plate works by interference and a Fresnel lens by refraction. The focal length of a Fresnel lens is not proportional to the wavelength of radiation, but for a zone plate, it is.

Different from the conventional lens, which uses curved surfaces to refract radiation, the so-called diffractive lens is made of a transparent flat plate with numerous zones; each zone has its location and optical characteristics precisely and, in common case, numerically defined. When radiation passes through these zones, the radiation waves interfere with each other due to phase differences and this results in radiation wave manipulation.

Fresnel came up with this idea in the early nineteenth century because he needed to construct a large lens with appropriate focal length for a lighthouse, but was unable to use the traditional shaped dielectric lens due to its excessive weight and size. Fresnel had reasoned that it was the surface curvature of the shaped dielectric lens, which gave it the focusing power; so, he instead reproduced the surface curvature in small sections. In this way, he was able to maintain the same focal length while reducing the weight and size of the lens significantly.

The Fresnel lens operation, like the shaped dielectric lens, depends on wave refraction and velocity changes in the material to perform proper focusing. Unfortunately, compared to the hyperbolic lens, the Fresnel lens is less efficient at focusing the energy from the source since the discontinuities in the aperture cause scattering and shadowing effects. Also, the Fresnel lens antenna has a more narrow band of operation since the bandwidth is inversely related to the number of zones. The efficiency and bandwidth are thus traded off against the improvement in size and weight.

1.3.3 Classical Fresnel Zones

The concept of Fresnel zones originated in the early nineteenth century from the work by Augustin Fresnel [9]. Fresnel expanded on Huygens' work by taking into account the space and time periodicity of light waves. He was interested in the interference and diffraction of light and was the first to use a geometrical construction of zones to solve what is now referred to as the Huygens-Fresnel integral [10]. Although these zones, which later became known as the "Fresnel zones", can be applied to different shaped surfaces [11, 12], the discussion in this work will be limited to a planar surface since it is the most appropriate for FZPL antenna.

Figure 1.7 illustrates the determination of Fresnel zones on a planar surface, which is normal to the direction of propagation, i.e., the z axis. A spherical wave emanates from a source, S, located a distance F from the planar surface, centered at O.

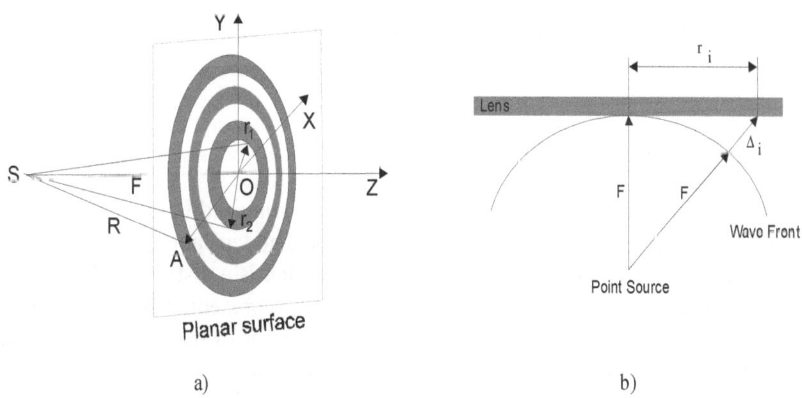

Fig. 1.7 Determination of Fresnel zones

The phase of the wave on the surface is what determines the zones. The phase at any given location on the surface can be calculated by first tracing rays from the source to the specific point on the surface. As shown in Fig. 1.7a, if the point on the surface is A, this creates a triangle between the rays SO, SA, and the radius r_A. The electrical path difference between SA and SO is the phase at point A. Since the source wave is spherical, all rays from the source to points on the surface at constant radii from O will have the same path difference. This results in constant phase spheres centered around O, which form circles on the planar surface. These circles represent the constant phase zones.

Mathematically, the phase at an arbitrary point on the surface, A, is computed according to Eq. (1.13), where F is the distance SO, R is the distance SA, r_A is the radius from O to A on the planar surface, and Δ is the difference in path lengths between SO and SA. The phase, ϕ, is calculated from the path difference via Eq. (1.14), where λ_0 is the free space wave length [12]:

$$\Delta = R - F = \sqrt{r_A^2 + F^2} - F, \tag{1.13}$$

$$\phi = 2\pi\Delta / \lambda_0 . \tag{1.14}$$

By knowing the phase values for every point on the surface, it is possible to define the zones. The first zone, from O to r_1 in Fig. 1.7a moving radially outward, contains the phases from 0° to 360°. The second zone, which covers the area between the first and second rings, r_1 to r_2, has phases between 360° and 720°, and so on. A complete zone, therefore, contains a 360° phase range.

It is also possible to define sub-zones where the phase from one sub-zone to the next ranges by an amount less than 360° [12, 13]. It is common to break the 360° phase range in half so that each sub-zone covers 180°. This is based on the half-period zone theory which Fresnel founded when he noted that light emerging from adjacent 180° zones will be opposite in phase.

1.3.4 Classical Fresnel Zone Radii

Simple design equations for the classical Fresnel zone radii are established based on the lens antenna geometry shown in Fig. 1.7b. The lens aperture has a diameter, D, and the source is located at the focal point, which is a distance F from the lens. The source generates spherical waves, which are represented in Fig. 1.7b by the hemispherical wavefront. The waves that hit the center of the lens have a different phase relative to the waves that hit the other parts of the lens. This is due to the varying path lengths from the source to the lens aperture.

This path difference was derived and expressed in Eq. (1.13). Re-arranging this equation and solving for the zone radii yields Eq. (1.15) where r_i represents the radius of an arbitrary zone and N is the total number of zones:

$$r_i = \sqrt{2iF\Delta + (i\Delta)^2}\ ,\ i = 1,2,3,...,N\,. \tag{1.15}$$

Equation (1.14) also needs to be re-written in order to account for the number of sub-zones, P, into which the lens is divided. The number of sub-zones is related to the phase by the following Eq. (1.14):

$$P = 2\pi/\phi,\ \text{Therefore}\ \Delta = \lambda_0/P\,. \tag{1.16}$$

where the phase, ϕ, is in radians.

Generally, as P gets larger, the lens will have more sub-zones and the phase error will decrease. In the common case, Eq. (1.15) may be re-written as follows:

$$r_i = \sqrt{(F + i\lambda_0/P)^2 - F^2}\,. \tag{1.17}$$

It can be easily shown that, if r_i /D is fixed and the focal length-to-diameter ratio (F/D) decreases, the normalized phase error, ϕ /D, will increase. This implies that, to maintain a given phase error at a fixed diameter with a decreasing focal length, more sub-zones are required.

1.3.5 Classical Circular Fresnel Zone Plate Lens Antenna Design

The FZPL antenna, as conceptualized by Soret and Rayleigh at optical frequencies according to [14], was first used as an antenna in the microwave band in the early twentieth century by Western Electric. In transmit mode, the FZPL antenna required that feed generating spherical waves be placed at the focal point of the lens so that a collimated beam would result in the far-field on the other side of the lens. The rings that were blackened by Soret to prevent light from getting through, were made metallic to reflect the microwaves. The resulting FZPL antenna, as illustrated in Fig. 1.8, consisted of planar circular rings, which alternated between transparent and opaque (metal) and became known as the circular FZPL (CFZPL) antenna.

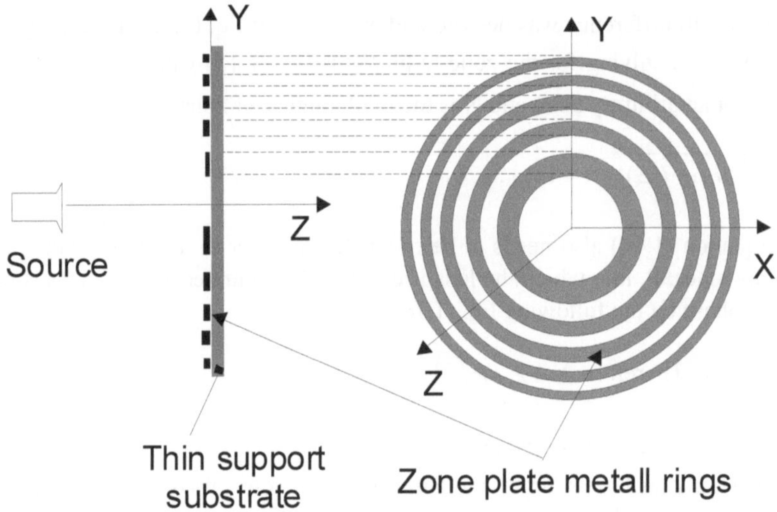

Fig. 1.8 CFZPL Antenna design: Side view (**a**), Front view (**b**)

The circular geometry stems from the fact that the spherical waves from the feed create constant phase zones on the planar surface that are circular. The metal rings coincide with the alternating 180° phase sub-zones on the surface of the antenna aperture. The EM waves that hit the transparent regions diffract through and combine to collimate a beam in the far field. This operation mechanism is different from the refraction and wave velocity method of the Fresnel and shaped dielectric lenses.

An interesting property of the CFZPL antenna is that in the first approximation it does not matter whether the first sub-zone is transparent or opaque. The zone plate lens will perform the same in either case since it can focus both in the transmission and reflection modes.

1.3.6 Phase Correcting Fresnel Lens Antennas

As mentioned previously, the FZPL antenna has a relatively low aperture efficiency. This is caused primarily by the fact that half of the radiation is wasted when it is reflected off the aperture back to the source. Another factor contributing to the reduction in efficiency is that the other half of the radiation that diffracts through the transparent zones of the lens does not add perfectly in phase. An antenna that avoids the first problem and improves upon the second is called the phase correcting Fresnel lens (PCFL) antenna. It can significantly improve aperture efficiency by performing path length adjustments, or phase corrections, along the lens surface. The idea is to transform the phase at discrete intervals across the lens aperture in such a way as to better approximate the phase correcting ability of the hyperbolic lens surface.

In 1898, Robert W. Wood designed the first PCFL at optical frequencies by using half-period (180°) phase correction instead of the blacked-out zones as Soret has done [5]. Researchers at Bell Telephone Labs filed a patent for the first PCFL antenna in 1939 [14], which was designed at 30 GHz with material having a dielectric constant of 20. In 1948, phase correcting lenses at 24.6 GHz were investigated at the National Research Lab in Washington D.C. [14]. Despite this activity, the PCFL antenna did not become popular until 1960 when Sobel et al. [15] designed and tested the first PCFL antennas at millimeter wave frequencies. Their designs were quite novel and involved phase corrections smaller than 180° which had never been implemented before. This led to aperture efficiencies on the order of 50–60%.

Since that time, many researchers from around the world have been involved in studying the PCFL antenna. As a result, several methods have been established to create the phase adjustments including:

- grooves or steps that are cut into the dielectric,
- stacks of dielectrics with varying dielectric constants,
- multiple layers containing metal rings,
- perforations in the dielectric material,
- printed metallic rings on a substrate and
- diffractive photon crystal lens.

Each method has advantages and disadvantages but, in all cases, the increase in the aperture efficiency of the lens comes at the cost of a substantial increase in antenna thickness compared to that of the standard FZPL antenna. Some of these methods will be reviewed briefly in the following sections.

1.3.7 Grooved PCFL Antenna

The grooved 180° PCFL antenna is an extension of the standard FZPL antenna and has been investigated by many researchers [13, 16, 17, 18, 19]. The metal sub-zones of the FZPL antenna are replaced with sub-zones that allow the radiation to pass through and also introduce the required 180° phase shift with respect to the adjacent zone. These sub-zones are thus made by cutting circular grooves, with a fixed depth, s, into a slab of dielectric material at the appropriate Fresnel radii. The groove depth is important since when chosen correctly, it will cause the radiation passing through the groove to undergo an 180° phase shift. In this way, the radiation that hits the grooves will be corrected and will be in phase with the radiation that exits in the adjacent zone. The operation mechanism is thus refraction and velocity changes instead of diffraction, which is similar to the dielectric lenses discussed earlier.

The calculation for the groove depth, s, involves the assumption that the rays going through the lens are parallel instead of being refracted by the dielectric material. Equating the path difference between the ray that goes through the groove and the one that does not, to $\lambda_0/2$ (180° phase correction) yields Eq. (1.18) where ε_r is the relative dielectric constant and P is the number of sub-zones as defined above [19, 20]:

$$s = \lambda_0 / \left(P \left(\sqrt{\varepsilon_r} - 1 \right) \right). \tag{1.18}$$

The aperture efficiency of the 180° grooved PCFL antenna is typically on the order of 40% depending on the F/D and frequency. This represents a small improvement over the standard FZPL antenna. However, by dividing the half-period (180°) sub-zones into smaller sub-zones, each with specific phase correction, it is possible to achieve better aperture efficiencies. By making smaller sub-zones, the grooved PCFL antenna will better approximate the cross-section of a Fresnel lens of the same F/D. In [15] it was shown that a grooved 90° PCFL antenna has a focal intensity of only 1 dB below the same F/D Fresnel lens. Also, a grooved 90° PCFL antenna design shown in [19] at 30 GHz with F/D=1 and D=15.8 cm achieved a 43% aperture efficiency, which is a significant improvement over the FZPL antenna and much closer to the 60% achieved with the hyperbolic dielectric lens.

However, there is a limit to how much phase correction can be realistically achieved. The authors in [19] indicated that using a phase correction beyond quarter-wave (90°) would not be worth the extra fabrication complexity and lens thickness required.

1.3.8 Shadow Blockage

The primary disadvantage of the grooved PCFL antenna is shadow blockage, an effect that limits the aperture efficiency by creating shadow regions. Black and Wiltse [4, 11] considered screening of Fresnel zones in a high-aperture zone plate. The problem was later treated in detail by Petosa and Ittipiboon [19, 21]. The radiation that hits the vertical sections of the groove refracts through the lens and adds constructively on the other side. The radiation that hits the horizontal sections of the groove, however, refracts in a direction that does not allow it to add constructively when it emerges from the lens. Since not all of the incoming radiation add constructively, the aperture efficiency is reduced.

In [21], also derived was the overall shadow efficiency for a given grooved PCFL antenna. The shadow efficiency is also highly dependent on the F/D. As outlined in [19, 20], when the F/D decreases, the shadow efficiency gets worse and can be as low as 0.4 for the F/D=0.25 case as compared to 0.9 when the F/D=1.0. This is because, when the feed is closer to the lens, the radiation hits the grooves at a small angle (< 35°) with respect to the lens surface, which allows more opportunity for the rays to hit the horizontal sections. When the feed is further away, the radiation hits the lens with bigger angles (> 60°).

It was shown in [12, 22] that this effect can be effectively reduced for diffractive elements fabricated on an arbitrary curvilinear surface, but this is not a subject of the present book.

1.3.9 Mismatch Losses

Another disadvantage of the grooved PCFL antenna is mismatch loss. Like shadow blockage, it limits the aperture efficiency of the antenna but for entirely different reasons. The mismatch loss arises from the radiation that is reflected off the face of the antenna aperture, which is an air-dielectric interface. The amount of reflection depends on the lens thickness, dielectric constant, and angle of incidence of the incoming radiation.

It is difficult to calculate the mismatch loss due to the nonuniform thickness of the lens and the varying incidence angles of the incoming radiation across the aperture. In [19] these losses were investigated by simulating and testing antennas. To observe the effect of the lens thickness, they tested two grooved PCFL antennas at 30 GHz with F/D=0.5 and with overall thicknesses, t=33 mm and t=15 mm. They found that the thinner lens had higher gain and hence lower mismatch loss.

To quantify the effect of the dielectric constant, the same authors tested two grooved 180° PCFL antennas at 30 GHz with F/D=0.25 having constant thickness but different dielectric constants, ε_r=2.5 and 10. The antenna with the lower dielectric constant had deeper grooves than the one with the higher dielectric constant, which caused larger shadow blockage areas. The deeper grooves were due to the relationship between ε_r and groove depth from Eq. (1.18). The higher dielectric constant material had bigger reflections off the surface, which caused the mismatch losses to increase. It turned out that the higher dielectric constant was more desirable since the higher reflections could be used advantageously to reduce mismatch losses.

It could be noted also the following. As it was described in [12, 23] in the general case, if a diffractive antenna is fabricated on a surface of revolution, including a flat, one always has to deal with a technological substrate on which the necessary phase profile of the antenna is machined. The optical thickness of this substrate (or the part of the lens that supports the grooves) may be comparable with the height of phase profile, and the Fresnel zone radii will not coincide with zone radii calculated in the approximation of "thin" diffractive antenna. This technological substrate that supports the grooves may be considered as a "refractive layer" of the diffractive antenna, which makes it possible to treat this "layer" as an additional degree of freedom; by choosing its profile, it is possible to correct for the beam pattern of the element as a whole [12]. This was accomplished by appropriately selecting the thickness of that part of the lens that supports the grooves [24, 25]. It was shown in [24] that the gain of the grooved PCFL antenna varies significantly with the thickness of that part of the lens that supports the grooves.

Figure 1.9 shows the focusing gain, G, as a function of the base thickness, t, for the two zone plates: F/λ=15 (solid line) and F/λ=20 (dashed line). For the figure, the wavelength in the dielectric is $\lambda_d = \lambda/\sqrt{\varepsilon_r} = 0.629\lambda$, and $0 \le t/\lambda \le 0.629$, or $0 \le t/\lambda_d \le 1$. The zone plate with the focal length of F/λ=15 has higher focusing gain when $t = 0.1\lambda$. Setting the base thickness to $t = 0.3\lambda$, however, would reverse the result, and the zone plate with F/λ=15 would outperform the zone plate with F/λ=20 by more than 1 dB.

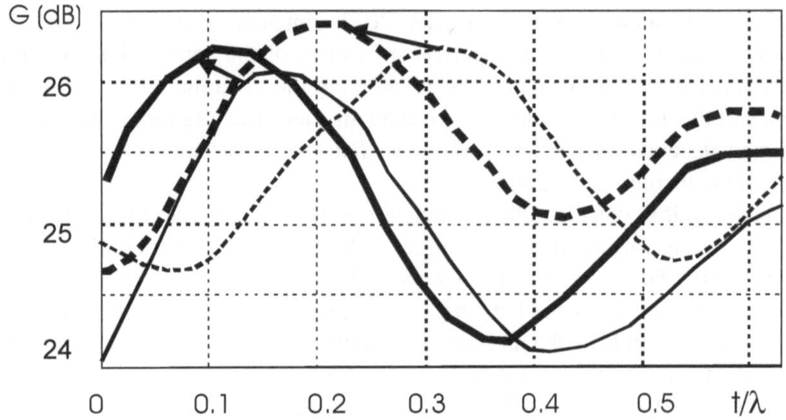

Fig. 1.9 Focusing gain, G, as the base thickness t is varied for two similar quarterwave zone plates. The solid line shows G for a zone plate with F/λ = 15 and the dashed line for a zone plate with F/λ=20. Both zone plates share the parameters N = 5; ε_r = 2:53, and P= 4. Thin lines are the zone plates without matching layer, thick lines are the zone plates with matching layer (after [26] with permission by Prof. Glenn S. Smith)

Matching layers are often placed on the surfaces of conventional lenses to reduce reflections and improve their performance, and we might expect the same to be true for the zone plate. To test this hypothesis, a dielectric, quarter-wave matching layer was placed on the front surface of the zone plate. The thickness of the layer was $(\lambda/\sqrt{\varepsilon_r})/4$, where, for this case, the relative permittivity of the layer was $\sqrt{\varepsilon_r} \approx 1.59$.

Figure 1.9 shows the effect of including the matching layer. It clearly alters the dependence of the focusing gain on the base thickness t/λ. For some values of t/λ, the layer increases G as expected, while for other values of t/λ, it decreases G. When t/λ=0.1, the matching layer increases G for both designs F/λ. However, when t/λ=0.3, the matching layer decreases G for both designs F/λ. Figure 1.9 also showed that the matching layer increases G and decreases t for both designs F/λ (see arrows in the Fig. 1.9). From these results, it appears that the effectiveness of a simple quarter-wave matching layer for increasing the focusing gain of the zone plate must be evaluated for each design.

The base thickness controls the overall thickness of the zone plate and, as shown above, affects the focusing ability of the zone plate. Despite this, it has received a relatively small amount of attention in the literature [24]. In a practical design, a thickness $t > 0$ is necessary to support the zone plate. In an analysis of these devices that use geometrical optics, this factor does not affect the focusing, because it imparts a uniform phase shift for all of the incident field. In reality, however, it does determine how much of the incident energy is reflected from the dielectric-air interfaces and how the multiple reflections that occur within the zone plate produce constructive or destructive interference.

Therefore, it is reasonable to assume that t can be optimized. An in-depth study, however, showed that there does not appear to be a simple, analytical method for optimizing t. The study also showed that for typical microwave or millimeter-wave zone plates, the change in the focusing gain due to varying t can be as much as 2 dB [26].

Let us consider the optimal value of P – a number of phase corrections per zone for minimizing reflection from the zone plate surface. The thickness of the dielectric for consecutive sub-zones decreases by the distance s. This distance is chosen to make the phase for a horizontal path through the zone plate decrease by at the start of each sub-zone within a full-wave zone (1.18).

If we consider the next sub-zones and write a condition of antiphase of reflected waves from the next sub-zone, we may conclude that:

$$2s \approx \lambda / 2 .$$

Keeping in mind (1.18), it is followed that:

$$P_{opt} \approx \frac{4}{\sqrt{\varepsilon_r} - 1} .$$

Taking into account that in microwave and millimeter wave bands, it is usually that $\sqrt{\varepsilon_r} \approx 1.5 \ldots 1.7$, it can be arrived at that the value of $P_{opt} \approx 8$.

1.3.10 Reflector-Backed Fresnel Zone Plate Antenna

The conventional FZPA has two disadvantages. The first, as it was mentioned earlier, is the inherently low aperture efficiency, which is a result of the significant amount of EM waves that are blocked by the metal zones. The second disadvantage is the large overall volume of the antenna due to the location of the feed relative to the aperture. The reflecting FZPA still suffers from the same large volume as the conventional version and requires an additional support for the feed.

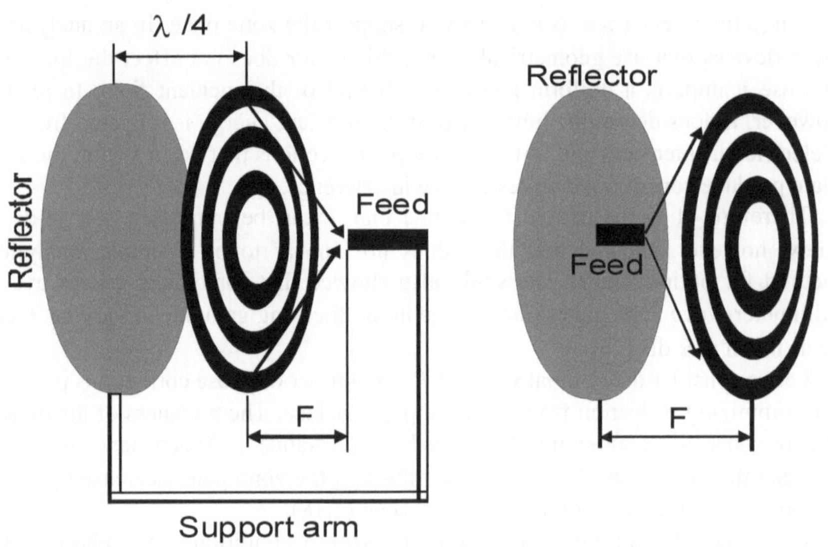

Fig. 1.10 FZPA in Reflection Mode (*left*) and Transmission Mode with Reflector (*right*)

In the paper [27], an alternative feed method was investigated, which enhances the aperture efficiency of the FZPA in transmission mode while allowing for a potentially significant reduction in profile. Instead of placing a reflecting surface behind the FZPA, it was placed at the aperture of the feed antenna as shown in Fig. 1.10. With this geometry, the FZPA could be fed using a planar antenna such as a microstrip patch. In this way, if the focal distance was decreased to about 0.5λ, the feed, a layer of foam, and the lens could all be sandwiched together to yield an attractive compact package, which would be comparable to a planar array in profile.

The FZPA used in [27] consisted of 10 zones. The focal distance-to-diameter ratio (F/D) was set at 0.237 and the focal distance was 3.75λ. The FZPAs were designed and modeled at 30 GHz with perfect conductors for the metal zones and reflecting surface. An open-ended WR28 waveguide with a ground plane around the aperture was used to feed the FZPAs and was placed at the lens' focal point. A waveguide feed was chosen in this study to simplify the model and focus on the effect of the ground plane. The ground plane size was set at 15.8 × 15.8 cm square, where each side corresponded to the diameter of the outer zone of the FZPA. The waveguide was oriented such that the electric field vector was along the y-axis.

Figure 1.11 shows FDTD-simulated comparison at 30 GHz of the H-Plane radiation patterns between two identical 10-zone FZPAs, where one has the ground plane at the aperture of the waveguide feed and the other does not have the ground plane. The addition of the ground plane improved the peak directivity at 30 GHz by about 1.3 dB and lowered the overall side lobe level by nearly 2 dB. As a result of the directivity improvement, the aperture efficiency improved from 12% to 16% when the ground plane was included.

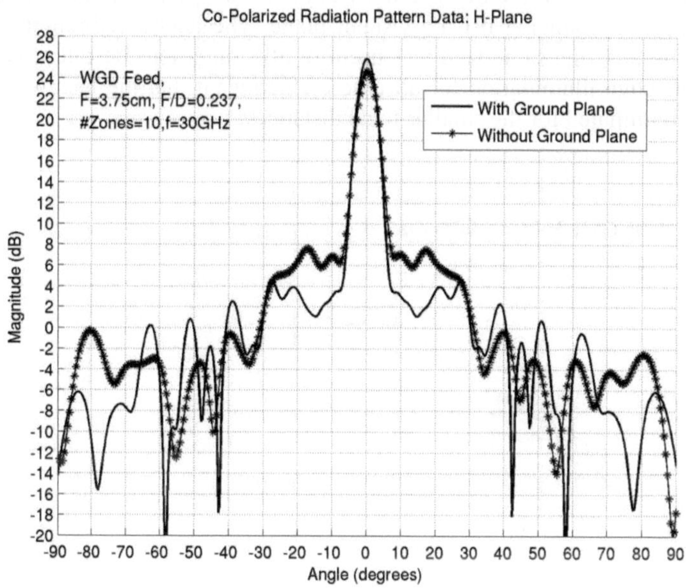

Fig. 1.11 Radiation Patterns of the 10-Zone FZPA with and without the Ground Plane: H-Plane

The reason for this behavior was related to the interference of the re-scattered fields from the ground plane. In other words, the fields that were initially reflected off the metal zones of the FZPA were being re-reflected by the ground plane and then either diffracted through the air zones or reflected again off the metal zones. The re-reflected fields that diffracted through the air zones contributed to the increase in peak directivity and the re-reflected fields that reflected again off the metal zones caused further interference of fields between the lens and ground plane. Measurements confirmed that the FZPA with a reflector, or ground plane, at the aperture of the feed had a higher directivity by about 1.3 dB, better overall side lobes by about 2 dB, and an improved aperture efficiency of about 4%.

It is possible to decrease the large volume of the Reflector-Backed FZPA. Several lower profile configurations were investigated in [28]. The focal distance of the FZPA was progressively decreased and the effect on the antenna's radiation patterns, peak directivity, and aperture efficiency was observed. The focal distance of the FZPA was decreased from 3.75 to 0.25 λ and, for each focal distance, the radii of the metal zones were recomputed. The intent of studying the effect of reducing the focal distance was to determine how small the focal distance could be before the radiation patterns deteriorated excessively. It was expected that the key factors contributing to the degradation would be the aperture illumination and the effect of fields bouncing off the ground plane. The aperture illumination was a problem because the same feed was used for all simulations despite it not providing 10 dB edge illumination for all cases. This meant that the lens would not be entirely illuminated for the smaller focal distances.

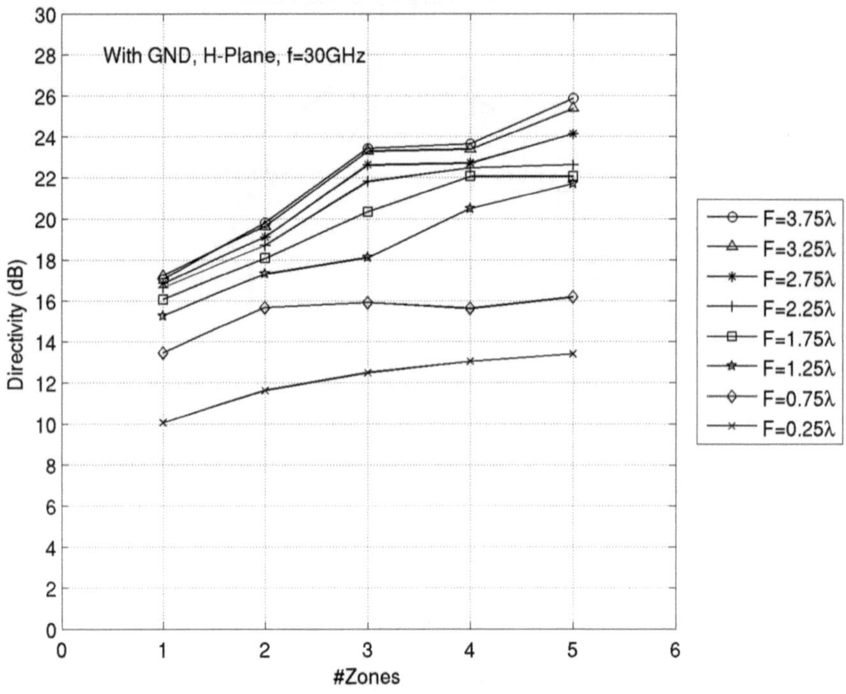

Fig. 1.12 Peak directivity vs. N of metal zones with ground plane

The effect of the number of metal zones on the peak directivity is shown in Fig. 1.12 for focal distances, which were decreasing by increments of 0.5 λ. The first observation was that the directivity decreased with decreasing focal distance. However, the cases with focal distances below 1.25 λ showed directivity decreases at 30 GHz, which were more than what was expected based on the aperture change. This additional drop was related to the aperture illumination effect since very little of the aperture was being illuminated at these small focal distances. When the focal distance spacing reached 0.25 λ, the directivity was only slightly higher than that of the feed. At these small focal distances, the near field or evanescent components of the field are involved, which means that the standard interference techniques do not apply [29]. In these cases, the lens tends to behave more like a leakywave structure. The other effect from Fig. 1.12 was that the directivity increased as the number of metal zones increased for nearly all focal distance cases. This indicated that the aperture illumination reduction with the larger number of metal zones was likely compensated for by the reflected fields from the groundplane as the number of metal zones increased. These groundplane reflections were evident in that the curves did not increase monotonically with the number of metal zones. When the same curves were generated without the groundplane, they increased monotonically.

Operating below 0.75 λ does not yield much more directivity than the feed, while operating at 2.75 λ produces nearly the same directivity as in the 3.25 λ and 3.75 λ instances. Also, it was found that the frequency where the maximum peak directivity occurred was not the same for each focal distance. Certain focal distances were, therefore, more optimal than others for focusing the fields at certain frequencies. This shift in frequency confirmed that the zone radii were not the proper sizes to obtain the peak at the designed frequency of 30 GHz.

Figure 1.13 shows how the aperture efficiency generally decreased with the increasing number of metal zones. The aperture efficiencies with five metal zones were much lower than with the single zone. This trend was the opposite of the directivity variation from Fig. 1.12. The reason that the aperture efficiency decreased and the directivity increased when the number of zones was increased is that the aperture area of the antenna increases faster than the directivity as the number of zones was increased. This makes sense in terms of the inverse relationship between aperture efficiency, physical area, and directivity.

Another observation was that for small focal distances, the aperture efficiency decreased nearly monotonically while this was not the case for the larger focal distances. Considering all the focal distances and zones, the highest aperture efficiency occurred for the F=1.25 λ case with the single metal zone, which was 66% higher than the single zone F=3.75 λ case.

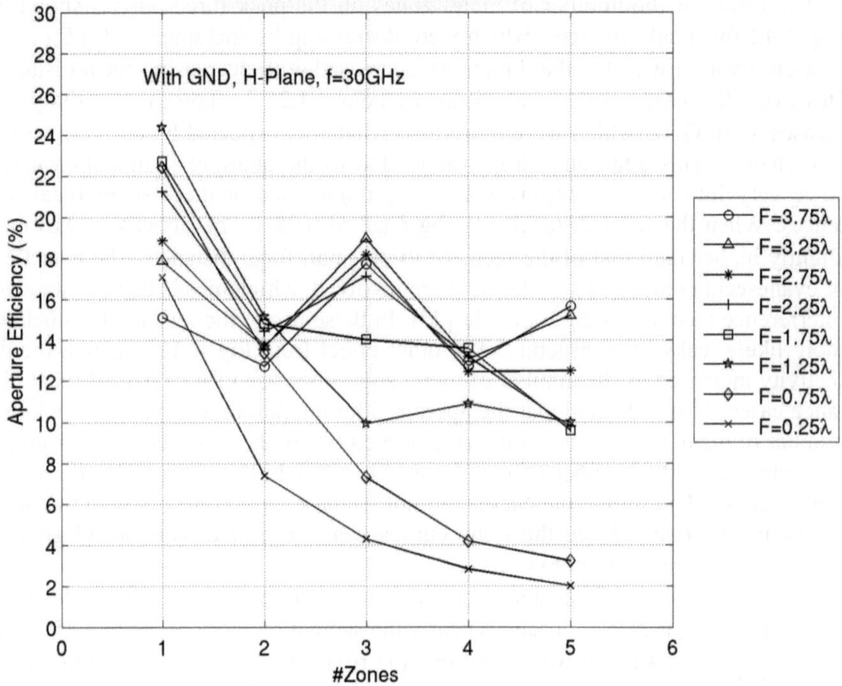

Fig. 1.13 Aperture efficiency vs. number of metal zones

It can be noted that an intriguing concept for alleviating these problems is the use of a three dimensional photonic crystal as the reflector.

1.3.11 Perforated Dielectric PCFL Antennas

The perforated dielectric PCFL antenna [19, 30, 31] is an extension of the multiple dielectric constant version. In this case, the multiple dielectric constants are achieved by perforating holes into a dielectric slab, which has a constant thickness and permittivity as shown in Fig. 1.14. The idea was derived from similar work performed on microstrip antennas in the late 1990s to enhance their efficiency and radiation characteristics [32, 33, 34, 35] and from similar work on perforated plate lenses [36]. From a fabrication point of view, drilling holes to control the dielectric constant is much simpler than machining rings out of different materials and trying to stick them together precisely.

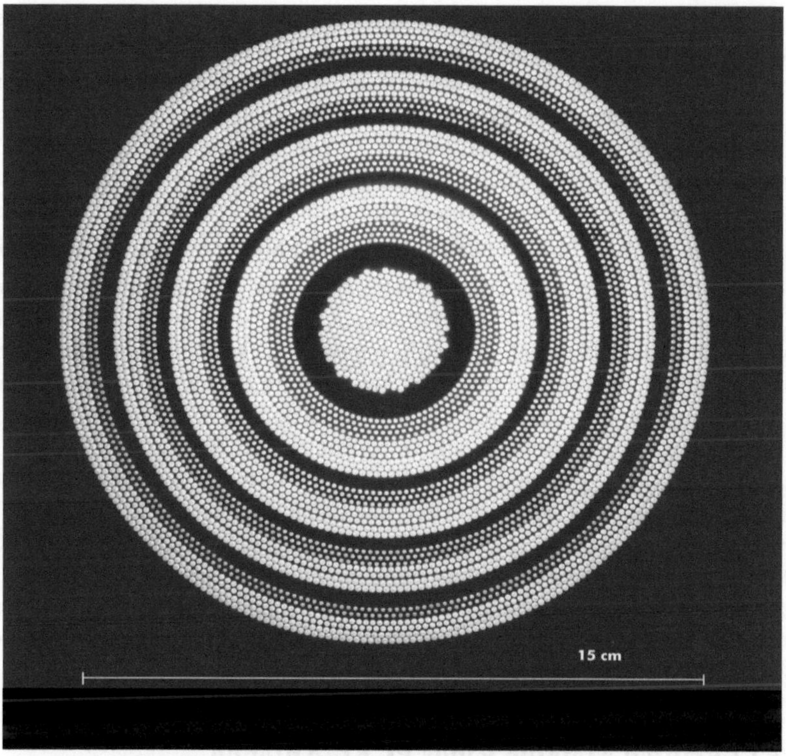

Fig. 1.14 Perforated FZP antenna (after [19] with permission by Dr. A. Petosa, Courtesy of the Communications Research Centre, Canada)

Perforating a dielectric substrate (ε_r>1) alters its permittivity because a portion of the original dielectric is removed and replaced with air which has ε_r=1. The permittivity of the substrate is, therefore, considered to be a combination of the air and the original permittivity thus creating an effective dielectric constant. By varying the size and spacing of the holes, it is possible to vary the effective dielectric constant through a range of values. The bigger the hole, the more air will be present, which will cause the effective dielectric constant to decrease. The range of dielectric constants is limited; however, if the separations between the holes in this lattice is kept small compared to the operating wavelength, the substrate will appear to have a uniform effective relative permittivity. This means that both the spacing between holes and the hole diameters must be small compared to the operating wavelength [19].

The arrangement of holes in the substrate also has a bearing on the effective dielectric constant. Typically, triangular and square hole arrangements are used. The important difference between the two arrangements is the area between holes, which has a strong bearing on the amount of holes that can be packed together, also called the filling factor.

It can be shown that the triangular lattice has the bigger fill factor and thus will enable a larger range of effective dielectric constants to be achieved. Using this lattice arrangement in [37], the authors designed a quarter-wave PCFL antenna with aperture efficiency of 35% at 30 GHz with F/D=1 and D=15 cm. This lens required 8800 holes spread over 7 zones. In [19, 38], the same authors looked at two perforated dielectric PCFL antennas having different dielectric constants, ε_r=5.8 and 10 respectively, in comparison with a grooved PCFL antenna having ε_r=2.6. The grooved PCFL antenna achieved the highest gain of the three with an aperture efficiency of 45%. The aperture efficiency of the perforated dielectric PCFL antennas was much lower and closer to 33%. This was attributed to the higher reflections off the air-dielectric interface because of the higher dielectric constants of the substrates in this case. The aperture efficiency could be improved by decreasing the dielectric constant, but that would cause the overall lens thickness to increase. Some useful details on perforated FZP may be found in [12, 19].

Kadri et al. in [39] expanded on this work by designing a multi-frequency perforated dielectric PCFL antenna. They found that careful selection of the hole diameter and location could enable the antenna to operate simultaneously at 20 and 30 GHz, the up-link and down-link frequencies for EHF SATCOM. They used a triangular hole lattice to design a 12-zone perforated dielectric PCFL antenna where the dielectric constants and filling factors were chosen so that they would converge for both frequencies.

1.3.12 Printed Metallic Rings PCFL Antenna

Guo and Barton [40, 41] designed a PCFL antenna using an array of printed metallic rings on a grounded substrate. Due to the presence of the ground plane, the antenna is in reflection mode. It achieves phase correction by adjusting the size and location of the planar conducting rings on the surface of the substrate.

To determine the appropriate phase shifts, they used the knowledge that small rings exhibited capacitive behavior while larger rings were more inductive. The antenna in [40] was designed to have quarter-wave phase correction at 10 GHz with only one full-wave zone and F/D=1.4 with D=33.2 cm. The antenna achieved 3 dB more gain than the reflecting FZPL antenna.

In [41], the same authors designed a larger version of the PCFL antenna, which covered three full-wave zones instead of the single zone from [40]. The diameter was increased to 59.4 cm which changed the F/D to 0.8 and the frequency was changed to 11 GHz. They achieved an aperture efficiency of 43%.

A single-layer ring reflector using circular, annular, or combined patches placed in classical Fresnel zones was described in [42, 43]. It could be noted that a quarter-wavelength substrate was used to simplify the design. The aperture radiation efficiency at 94 GHz was 34%. Two zonal ring reflectors were made, measured, and compared with the classical half-wave Fresnel zone plate reflector showing an improvement of 50% aperture radiation efficiency for the smallest one (four correcting Fresnel zones) and 70% for the largest (combined eight and four correcting Fresnel zones). Three types of patch elements were used: circular patch element, annular patch element, and combined circular and annular patch element. Due to the use of circular shapes of patches, i.e., circular, ring, and combined circular and ring patches, associated to reduce cell size, phase compensation in ($\lambda_0/8$) Fresnel zones in the reflector center was achieved. The ring and combined ring and circular patches shapes also insure an increase of the bandwidth while keeping a single-layer structure in comparison to classical solutions based on the use of multi-layer structures. Figure 1.15 illustrates an example of such an FZP antenna.

Fig. 1.15 Ring-reflector FZP antenna (By courtesy of C. Migliaccio et al., LEAT, UNS-CNRS)

1.3.13 Subwavelength FZP

Refractive optics, Fig. 1.16a, can provide ideal phase profiles but are nonplanar and not easily fabricated. The Fresnel zone equivalent, Fig. 1.16b, is almost planar reducing the refractive phase profile to a modulo 2π stepped phase profile. Some progress has been made in demonstrating these structures, but the fabrication is challenging. Conventional, multi-level binary FZPs, Fig. 1.16c, are planar, but only approximate the smooth phase profile of the Fresnel zones. The gradient index structure, Fig. 1.16d, is perfectly planar and provides the desired phase profile, but is practically difficult in this ideal form. It can be mentioned that the following design of FZP is possible. The lens is divided into sub-zones, each zone with linear-varying permittivity. The design is such that the phase at the output plane of the lens is equalized for all rays through the zone boundaries. Investigations showed that aperture efficiency of 81% is real.

Diffractive lens with feature sizes smaller than the wavelength of illumination are called subwavelength diffractive lens (SWDLs) – Fig. 1.16e. Because of their small feature size, SWDLs can be designed to perform the same function as multi-level lenses. When an incident wave approaches the subwavelength grating grooves, diffractive, rather than refractive, optics dominate. The effective index of refraction encountered by the incident wave depends not only on the substrate, but also on the spacing of the grooves and the relative amount of material with respect to air that the wave comes in contact with. The effective index of refraction seen by the incident wave is therefore a weighted average of the refractive index of the material and that of air. By judiciously selecting the feature sizes and their spacing, one can design a subwavelength lens to implement the same effects of phase control and constructive interference as with a multi-level lens and focus radiation accordingly, with little loss in performance [44, 45].

The simplest approximate model for subwavelength gratings is classical form birefringence [46]. This zero order approximation gives the effective refractive indices of a binary subwavelength grating as:

$$n_{TE}^2 \approx qn_1^2 + (1-q)n_2^2, \quad n_{TM}^2 \approx (n_1 n_2)^2 / \left[qn_2^2 + (1-q)n_1^2 \right], \tag{1.19}$$

where TE denotes light polarized parallel to the grating stripes, TM denotes light polarized perpendicular to the grating stripes, n_1 and n_2 denote the refractive indices of the materials that comprise the grating, and $q=t_1/\lambda$ is the duty cycle of the grating, i.e., the relative portion of the material with refractive index n_l within the grating. Note that the period is smaller than the incident wavelength λ. If the grating period is not binary, then it is approximated with a step-function, and the effective refractive indices for each step are calculated using Eq. (1.19). The structure is then replaced with a multi-layer stack whose properties can be calculated using transfer matrix methods [47, 48].

In addition to their focusing capabilities, SWDLs can also function as effective linear polarizers. Their inherently strong polarization dependence results from the particular direction of the subwavelength gratings, which favors one linear polarization of radiation over its orthogonal component.

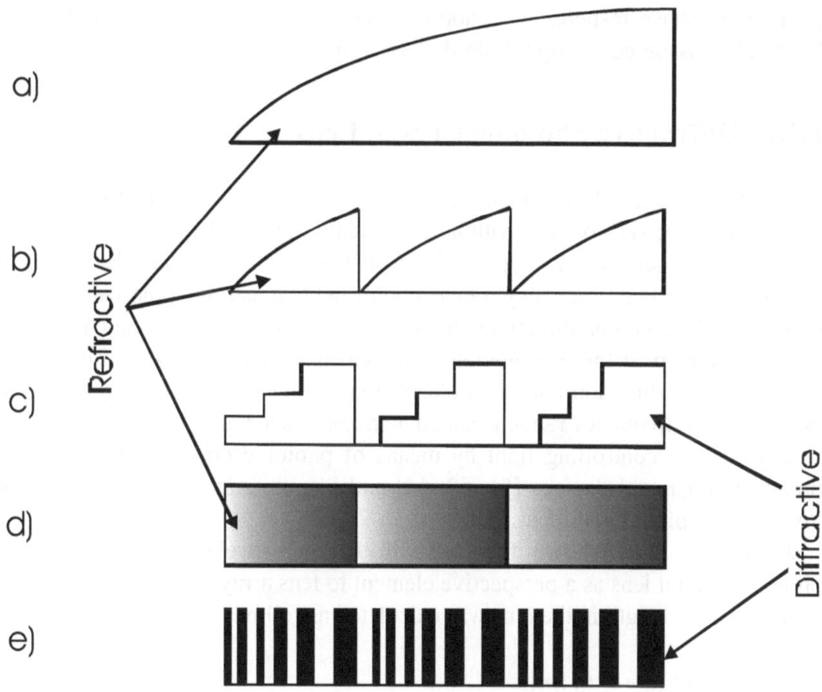

Fig. 1.16 Four main types of lenses: Segment of an arbitrary refractive lens (**a**), Fresnel lens (**b**), four-level Fresnel zone plates (**c**), gradient-index lens (**d**) and subwavelength artificial gradient index lens (**e**)

1.3.14 Perforated Subwavelength Diffractive Lens

Most diffractive elements (DEs) are designed using algorithms based on scalar diffraction theory and are fabricated with large features relative to the operational wavelength. Unfortunately, at longer wavelengths, the size of scalar domain features tends to be sufficiently large as to prohibit the fabrication of a practical DE that meets size and weight constraints. In [12, 49], a design algorithm was presented that utilized a look-up table of effective properties as a function of the grating's parameters as building blocks to synthesize more complicated DEs. The subwavelength cell-encoding procedure described in [12] was extended in [50] to include subwavelength features that are cylindrically shaped. The advantage of using cylindrically shaped subwavelength features is that inexpensive fabrication processes such as conventional mechanical drilling can be employed for DE fabrication. The authors [50] also extend the concept of subwavelength DEs to longer millimeter wavelengths. To design DEs with spatially varying effective properties, a cell-encoding scheme was employed. In this process, a DE with a desired continuous phase response was approximated by a number of finite size cells. Each cell is assigned a single desired phase value.

1.3.15 Diffractive Photonic Crystal Lens

One of the complex electromagnetic systems of interest to us is a photonic crystal, a periodic dielectric structure with lattice spacing of the order of the wavelength of the electromagnetic wave. Typical for a photonic crystal is that electromagnetic waves in a certain frequency range and/or with a certain polarization cannot propagate along certain directions in the crystal. This forbidden frequency range is called a stopgap. If the propagation of the electromagnetic wave is forbidden for any crystalline direction and any polarization, for a certain frequency range, then this forbidden frequency range is called a photonic band gap [51, 52].

The idea of controlling light by means of photonic crystals has led to many proposals for novel devices [51, 53, 54] and has motivated many researchers to investigate a plethora of ideas.

Below, we have shown the possibility of subwavelength focus diffractive photonic crystal lens as a perspective element to lens array.

The optical length in the lens was calculated as follows:

$$\Delta = N[2r_i(y) + (a - 2r_i(y))n],$$

Where, N – the number of holes in line, a – array parameter, and $r_i(y)$ – the radius of holes. From the equation, it is observed that at least 3 different types of photonic crystal lens are possible.

Let us consider the first type of lens where N=constant. The main parameters of diffractive photonic crystal lens are the following:

- Wavelength $\lambda = 10$ mm,
- Width of a lens $l = (0.83+1)\lambda = 18.3\,\text{mm}$
- Array parameter $a = l/10 = 1.83\,\text{mm}$
- Radii of a circular holes: $r_1 = 0.25a = 0.457\,\text{mm}$,
  $r_2 = r_1 + 0.227a = 0.87\,\text{mm}$
- Lens positions at z-axis: 15–33 mm
- Index of refraction n=1.6

The main results of FDTD simulation are shown in the Figs. 1.17 and 1.19 below.

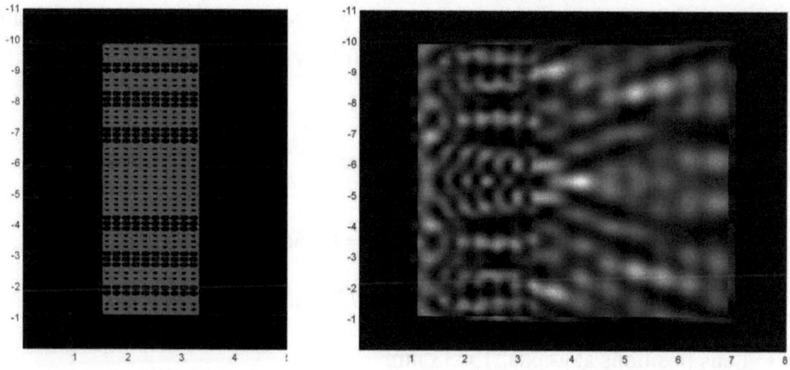

Fig. 1.17 Diffractive photon-crystal lens of the first type (*left*) and field intensity distribution along optical axis (*right*)

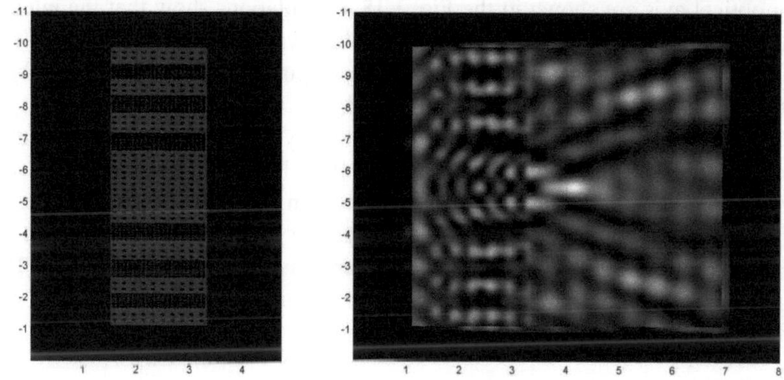

Fig. 1.18 Diffractive photon-crystal lens of second type (*left*) and field intensity distribution along optical axis (*right*)

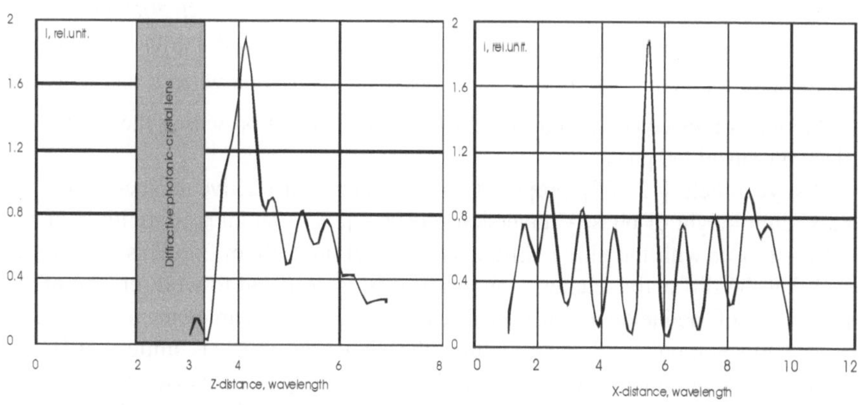

Fig. 1.19 Field intensity distribution along (left) and across ("focal spot", *right*) optical axis

For the lens of second type where $r_1 = r_2$ and N$\neq$ constant, the main parameters of diffractive photon-crystal lens are the following:

- Wavelength $\lambda = 10$ mm,
- Width of a lens $l = (0.83+1)\lambda = 18.3\text{mm}$
- Array parameter $a = l/10 = 1.83\,\text{mm}$
- Radii of circular holes: $r = qa_1 = 0.2a = 0.366$ mm,
- Array constant $a_2 = a_1 / \left(1+\dfrac{n-1}{2q}\right) = 0.732$ mm
- Lens positions at z-axis: 15–33 mm
- Index of refraction n=1.6

Diffractive photon-crystal lens of second type and field intensity distribution along optical axis are shown in the Fig. 1.18. Calculations show that the intensity in the focus increases by 12% and the side lobe level decreases up to 15%. It can be noted that the described results show only a possibility of diffractive photon crystal lens realization and could be optimized.

Thus, we applied the principles of photonic crystal devices to the millimeter wave portion of the electromagnetic spectrum [99]. For the lens, we have observed their collimation and imaging ability both shown in the amplitude and phase. The results described above show that the diffractive photon-crystal lens is a perspective candidate to subwavelength focus lens as an array element. The FDTD simulations of diffractive photon-crystal lens were made in cooperation with Prof. V. Kotlyar, Samara, Russia.

1.3.16 Phase Compensation for FZP Antenna of Big Diameter

It can be derived from Eq. (1.20) that the space dedicated to each zone defined by $(r_n - r_{n-1})$ is decreasing when n increases. This is a drawback because it might be impossible for example to obtain $\lambda_0/8$ phase compensation when n increases due to the lack of space compared to the size of the element providing the desired reflected/transmitted phase.

Malyuzhinets [12, 55] proposed to fabricate a diffractive element with the maximum height of phase – inversion profile equal to an integral multiple of $h = \lambda/(n-1)$ - or with the phase shift as an integral multiple of 2π. This signifies in fact that the diffractive element, or a part of it, is designed to work on one of the harmonics of the incident radiation. The advantage of such an element is a considerably lower number of zones within the aperture, which is important in the short-wavelength range. A diffractive element with a phase step equal to $h=Nh_0$ is calculated in such a way that the first several zones are phase-shifted by 2π, the subsequent several zones by 4π and so forth, up to the phase shift of $2\pi N$. The number of divisions for each value of the phase shift was equal for each group, even though this is not mandatory. Therefore, another possible free parameter (a degree of freedom) is available for optimizing the properties of diffractive elements: the height of the phase-inversion step. Furthermore, the frequency band with such a diffractive element also increases. This conclusion follows from the argument that the phase profile height for several wavelengths in the spectrum will be a multiple of 2π. Zone radii of the diffractive element are found from the expression:

$$r_i = \sqrt{2i\lambda_M MF + i^2\lambda_M^2 M^2}\,, \tag{1.20}$$

where M is the multiplicity coefficient for the wavelength λ_M. It is clear from this formula that the diffractive element computed for certain λ_M and M will have the same focal length for other wavelengths for which the product $\lambda_M\, M$ remains constant. However, in such types of FZP, the thickness increases.

The dielectric Fresnel lens was shown to have the best radiation performance and highest aperture efficiency (~50%). The FZPL antenna, on the other hand, has an aperture efficiency of less than 10% but was found to be the simplest to fabricate and of the lowest profile of all the lens antennas considered. The RFZPL, DFZPL, SWDL, and PCFL antennas, were developed as alternatives to the FZPL antenna with improved aperture efficiencies, which in certain cases, approach that of the dielectric Fresnel lens. Unfortunately, in all cases, the increase in aperture efficiency comes at the expense of increased cost, complexity, and size of the antenna.

1.4 Reference Phase Concept

The problem of synthesis of the amplitude-phase-type profile does not have a unique solution even for the simplest element – the Fresnel zone plate. In defining these zones of constructive and destructive radiation, Fresnel [9] assigned a specific phase origin or reference phase of 0° to the shortest ray connecting the source and detection point in order to simplify the analysis. Based on Fresnel's analysis, Soret later demonstrated focusing in half opaque zone plates, which blocked radiation from what was defined as the destructively interfering zones [56]; the same choice of phase origin as Fresnel was used. Later work has shown that the out-of-phase zones need not be blocked if selective phase shifts are introduced into the zone plate [14, 57, 58]. However, the same special choice of phase origin as Fresnel has been made.

1.4.1 Diffracting Grating and Fresnel Zone Plate

Let us consider two parallel rays incident at an angle θ_0 and striking two adjacent slits, of separation, a, at the same relative location. When the path length difference between the emerging rays is an integral number of wavelengths, constructive interference occurs and we get an interference maximum in the transmitted pattern. That is:

$$a(\sin\theta_n + \sin\theta_0) = n\lambda, \tag{1.21}$$

where n is any integer known as the order and θ_n is the corresponding angle. Both n and θ_n can take on positive or negative values, depending on whether the rays bend down or up.

Although we can pair off all rays from adjacent slits in the above manner for an interference maximum, it may also happen that rays from within the same slit combine to give total destructive interference for the very same conditions. Thus, the expected interference maximum would be obscured or "suppressed" due to the single slit diffraction pattern. This is known as a missing order.

However, under illumination from a point source (A), the phase of the radiation across the grating will be varying as radiation further from the axis has a greater distance to travel. One can imagine generating a grating with a varying slit separation, d, to correct for this phase difference allowing for an image of the source (B) to be generated. Using simple geometry, we can see that:

$$\tan\theta_0 = R/A, \ \tan\theta_n = R/B.$$

We can then substitute this into Eq. (1.21) assuming paraxial rays so that $\tan\theta \approx \sin\vartheta$ to give:

$$\frac{1}{A} + \frac{1}{B} = \frac{n\lambda}{dR}, \tag{1.22}$$

where the constant slit separation, a, of Eq. (1.22) has been replaced by the varying width, d, used here. When the spacing varies inversely with the distance from the axis, the right hand side of Eq. (1.22) is constant. Note that we are presuming a grating with a slit width that is half the slit separation. Diffraction effects then tell us that every second order will be missing. Hence n takes on only odd values in the above equation.

FZP is a series of concentric circles on a flat plate with the alternate regions darkened. It is constructed by imagining that there is a radiation source of wavelength λ at a distance A away from the plane of the plate. At a given instant in time, the radiation on the axis of the plate will have a certain phase. However, radiation on other portions of the plate will have a different phase since it had to travel further. The circles on the plate, which mark the zone boundaries, are those places where the radiation is exactly in phase or exactly out of phase with the light on the axis. Let us treat the zone plate as a collection of infinitesimal linear gratings with a spacing, d, which varies in a known way with the radius R. We can see that:

$$d/2 = R_m - R_{m-1}.$$

On the other hand, $\tan\theta \approx \sin\theta = \lambda / d$, which shows that the slit separation varies inversely with radius. Substituting this expression into Eq. (1.22) yields the formula:

$$\frac{1}{A}+\frac{1}{B}=\frac{1}{f_n},\; f_n=\frac{F}{n}\left(\frac{\lambda}{\lambda_0}\right).$$

This last formula is recognized as being the familiar Gaussian lens formula for a lens with a focal length f_n. That is, the zone plate will act just like a thin lens with a focal length f_n. Also, since n can take on negative values, the zone plate acts like a diverging lens as well as a converging lens. Furthermore, since n can take on several integral values, it will behave like very many diverging and converging lenses all rolled into one. Notice also that the focal length is wavelength dependent.

1.4.2 Physics of Reference Phase

Let us consider FZP with N>>1, where N is a number of Fresnel zones. In this case FZP could be interpretive as a quasi-periodical grating.

As it is known in a circular grating (Fig. 1.20), as spatial overlapping of positive and negative orders of diffraction (*n*) with same number is observed. This fact is the reason for other specific phenomenon, which is describing a property of a circular grating: the radial structure of field intensity at diffractive maximum ($n \neq 0$) in the Fraunhoffer zone depends on the relative phases of wave fronts. Phases, on the other hand, are defined by position of a grating of radius R_0 at coordinate r (where the grating begins), and an initial phase from which the periodic transmitting function t(r) will start at the point $r = R_0$.

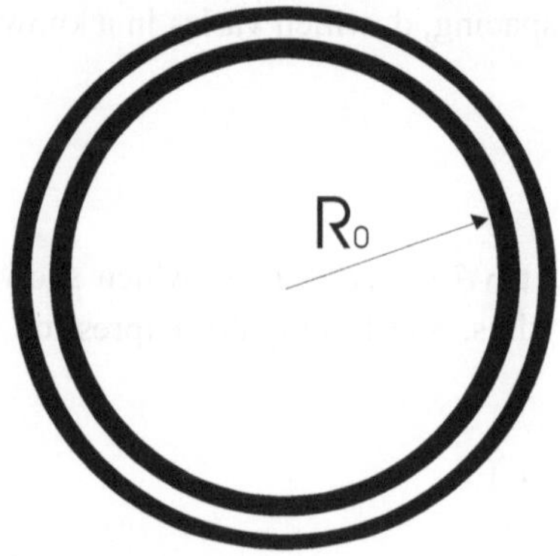

Fig. 1.20 Circular grating

Let us write down the expression of a spatial part of an electric field of a wave in Fraunhofer zone diffraction *E (r),* through a field of a wave at once behind grating *A (r)*. Let us analyze a circular aperture in the Fraunhofer approximation, namely a distant source and a distant observation point. Therefore, plane waves are incident on and leave the aperture. At circular symmetry, two dimensional Fourier transformation becomes one dimensional transformation of Fourier-Bessel

$$E(r',z) = \exp(\frac{ikr'^2}{2F})\int_0^\infty E(r)J_0(k_r r)rdr \cdot \qquad (1.23)$$

For simplification of a problem, we shall consider the elementary case of a harmonic grating, when:

$$E(r) = E_0 t(r) = 0.5E_0\left(1+\cos(\frac{2\pi r}{d}+\delta)\right)\cdot \qquad (1.24)$$

Note that d is independent of r, transmittance of the grating is 0.5 $(1+\cos((2\pi r/d)+\delta)$ and δ is the "reference phase" for the grating. There is a circular grating in the aperture with a variable transmittance, which depends on radius r in the aperture.

For further simplification, we shall replace Bessel function it asymptotic approach, which are fair both for all not zero orders of diffraction and at conditions of $2\pi nr/d >> 1$:

$$J_0(k_r r) \approx \sqrt{\frac{2}{\pi(k_r r)}}\cos\left(k_r r - \frac{\pi}{4}\right)\cdot \qquad (1.25)$$

After substitution of (1.24) and (1.25) in (1.23) and the lowering of high – oscillating members, it is obtained that:

$$E(r') = -0.5iE_0\sqrt{\frac{k}{2\pi r'F}}\exp\left(ik\left(z+\frac{r^2}{2F}\right)\right)\times$$
$$\int_0^R\left[\exp\left(i\left(\delta+\frac{\pi}{4}\right)\right)\exp\left(\frac{i2\pi r}{d}\left(1-\frac{rd}{\lambda F}\right)\right)+\exp\left(-i\left(\delta+\frac{\pi}{4}\right)\right)\exp\left(\frac{i2\pi r}{d}\left(1-\frac{rd}{\lambda F}\right)\right)\right]rdr \tag{1.26}$$

From (1.26), strong dependence of diffractive structure rings from an initial phase of a grating is visible. From (1.26), it also follows that intensity in the center of a diffractive maximum at $r' = \lambda F / d$ is:

$$I_r = \frac{4dR^3E_0^2}{9(\lambda F)^2}\cos^2(\delta - \pi/4)\cdot \tag{1.27}$$

For a linear one dimensional grating of the same radius R and with the same transmitting function $t(x)$, intensity in the center of diffractive maximum (which is the disk of Airy in this case) is given by the known expression:

$$I = \frac{R_0^4 I_0 \pi^2}{16(\lambda F)^2}\cdot \tag{1.28}$$

From the comparison of expressions (1.27) and (1.28), the dependence of the field intensity in the center of diffractive maximum vs initial phase of transmittance function of a circular grating is clear visible. Thus, the intensity in the center of diffractive maximum of a circular grating depends on the size of a grating. In case of a circular grating, the width of diffractive maximum is proportional to $\Delta r \approx 1/R$, but in so called Airy's disk its area is proportional to s~ R_0^2.

In the standard Fresnel zone plate, the cos term in (1.24) would be either 0 or 1 and d would not be constant due to phase requirements for constructive interference. Also taking into account δ as the reference phase for the grating, it is very analogous to the reference phase in Fresnel zone plates [12]. Again, it is interesting to note that if the number of Fresnel zones in the Fresnel zone plate is much more than 1, then it may be shown as a circular grating.

1.4.3 Geometrical-Optics Interpretation

As an example, we consider a diffractive element that allows focusing of radiation onto an annular region. It appears that a diffractive element focusing laser radiation onto a ring was first described in [59]. It was meant to produce annular imprints on metal surfaces. The field structure in the focal region of this element of diffractive optics was later treated in the Fresnel approximation in [60, 61]. Specifically, it was shown [60] that there are two solutions to the problem of focusing onto a ring: one in which rays intersect the optical axis and another in which they do not intersect it. A similar analysis of elements with angular momentum response was given in [62] in relation to systems of geometric transformation of images. In addition, it was shown in this paper that field intensity has a maximum in the focal plane on the optical axis of the element of diffractive optics, produced by weak, spurious illumination along its optical axis [62]. By increasing the number of phase quantization levels, *P*, of a phase element of diffractive optics approximately to 16, it is possible to greatly reduce the field intensity at the optical axis, as shown in [63, 64] in the paraxial approximation. However, in the optical range, this increases the technical difficulties in manufacturing the diffractive element and multiplies the number of masks required to generate the desired phase profile [63, 64]. Also, in a number of cases, there are reasons that make this approach (calling for increased number of phase quantization levels) unsuitable.

In addition to increasing the number of phase quantization levels, there is another free parameter of diffractive optical element whose selection can regulate the field intensity distribution function in a prescribed region of space [23, 65, 66].

To simplify the explanation of the method, we consider here how monochromatic radiation from a point-like source is focused to a point using a flat diffractive element. The structure of zones on its surface can be easily obtained by considering equiphase contours (Fig. 1.21a, b); these lines are elliptical.

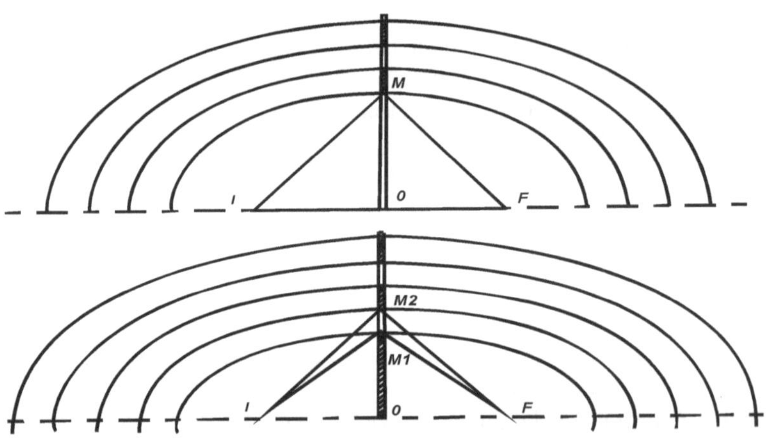

Fig. 1.21 (**a, b**). Principle of construction of equiphase curves (after [63]).

Obviously, to focus radiation at a point F, it is necessary (for a binary element) that the difference between the eikonals of the incident and diffracted rays, $|IOF|$ and $|IM + FM|$ be a multiple of $\lambda/2$ (see Fig. 1.21a). Since the phase shifts on the boundaries between the neighboring zones have opposite signs, transparent and opaque zones alternate. Practically all known binary diffractive focusing elements built until recently followed this principle. On the other hand, the central zone of a diffractive element can be ether transparent or opaque for the incident radiation. In the latter case, the in-phase condition for the radiation at the point F is written differently: the difference between the eikonals of the rays, $|IM_2 + M_2F|$ and $|IM_1 + M_1F|$ must be a multiple of $\lambda/2$ (see Fig. 3.1b). We see from Fig. 3.1 that in this case, the size R_0 of the central opaque zone $|OM_1|$ can be arbitrary. It is easily seen that it is expedient to choose $R_0 \leq |OM|$ since otherwise such effects as apodization will be observed. It is not difficult to apply similar reasoning to other types of diffractive focusing elements, such as focusing onto a ring.

In other words, another parameter that can be varied in calculating the modulation function of a diffractive element is a permanent phase shift. The diffraction zones can be displaced by adding an initial phase shift φ_0 to the phase function of the element:

$$\bar{\varphi}(r) = \mathrm{mod}(\varphi(x, y) + \varphi_0, 2\pi).$$

This means, in fact, that the zone is divided not by starting at the extremum of the transmission phase function, but at an arbitrary value of transmission (Fig. 1.22).

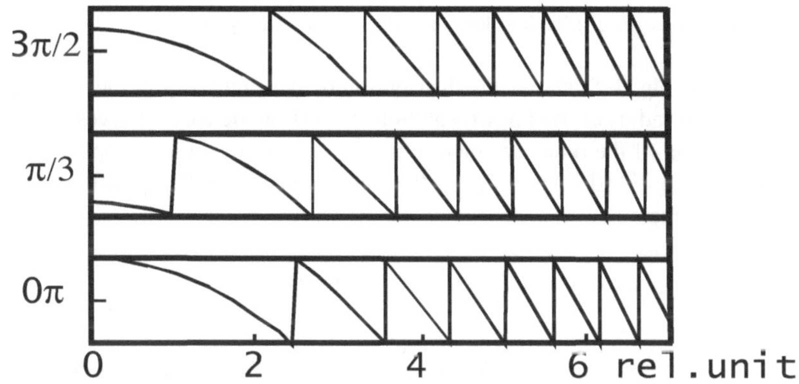

Fig. 1.22 Phase profile for various values of phase shift: 0π, π/3, 2π/3.

For a binary element that focuses radiation emitted by a point-like radiation source to a ring of radius R_{ring}, the effect of R_0 on the quality of field intensity distribution function in the focal plane was evaluated by running a numerical experiment in scalar wave approximation.

We chose the following parameters for the focusing element: distance A from the point source of radiation to the diffractive element plane and distance B from this element to the focal plane were 7.5 cm each, the diameter of the diffractive optic element D was 2 cm, the radius of the ring of focusing R_{ring} was 0.1D, and the wavelength λ was 10.6 μm. The results of the numerical experiment are plotted in Fig. 1.19 (the ordinate is the field intensity normalized to the largest maximum of the three distributions under study, and the abscissa axis is the coordinate along the ring radius in the focal plane). Figure 1.24 shows the corresponding shapes of focusing areas in isometric projection.

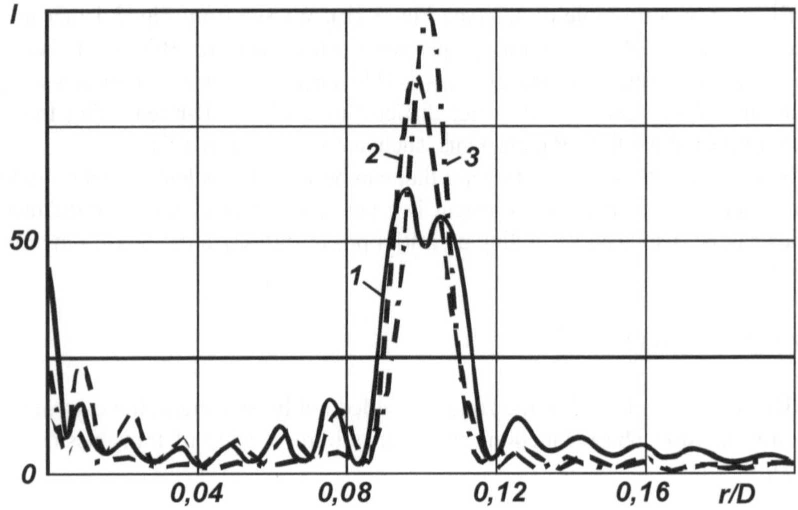

Fig. 1.23 Field intensity distribution along the radius of the focusing ring

Curve 1 in Fig. 1.23 shows the distribution of field intensity in the focal plane along the ring radius of the known solution ($R_0 = 0$) while curve 2 gives the result of optimization of the binary element with the central maximum suppressed. The optimization consisted in minimizing the field intensity on the optical axis of the diffractive element R_0:

$$R0: I \sim \min_{Ro} |U(x = 0, y = 0, B)U^*(x = 0, y = 0, B)|.$$

In all cases shown in Fig. 1.23, the central zone of the diffractive optical element was opaque for radiation and the total number of zones on the surface of the focusing element remained unchanged.

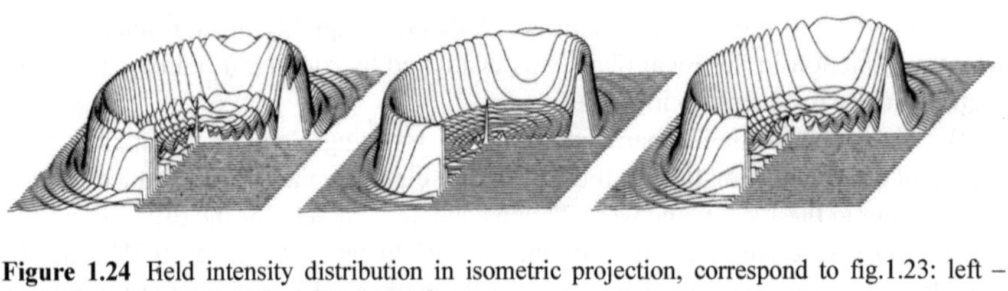

Figure 1.24 Field intensity distribution in isometric projection, correspond to fig.1.23: left – curve 1, average – curve 2, right – curve 3

As follows from Fig. 1.23, the stray central maximum of field intensity can be essentially "suppressed" by choosing a suitable value of R_0. The gist of the effect observed is as follows. The boundary of each zone on the surface of a diffractive element is a source of toroidal convergent boundary waves, so that in-phase constructive interference of these waves can be observed not only in a prescribed space (at a point R_k) but also in some different locality, for instance on the optical axis where the effect reaches its maximum. By redistributing the structure of zones on the surface of the diffractive element, the synchronization condition for boundary waves in a certain prescribed region can be changed by choosing the value of R_0. This makes it possible to considerably suppress the spurious central maximum even for a binary diffractive element and to increase field intensity in the ring itself through spatial redistribution of energy.

Similarly, it is possible to correct the shape of the focusing region. For instance, the problem posed by focusing radiation onto a ring with maximum intensity at a point R_k was solved in this manner, that is, through optimization of the type:

$$I \sim \min_{Ro} |U(R_k,0,B)U^*(R_k,0,B)|.$$

The results of optimizing the focusing properties of diffractive optical elements are shown in Fig. 1.23 (curve 3).

Some comparative characteristics of the discussed variants of diffractive optical elements are shown in Table 1.1.

Table 1.1

Curve in Fig. 1.19.	R_0[1]	I/I_{max}[2]	I_0/I[3], %	$\Delta r/R_{ring}$[4]		
				0.1	0.5	0.7
1	0.00	0.69	65	0.302	0.238	0.198
2	0.47	0.81	21	0.270	0.143	0.103
3	0.64	1.00	25.5	0.218	0.132	0.094

Note:
1. The value of R_0 is given in units of $(\lambda B)^{1/2}$.
2. I/I_{max} is the normalized field intensity at the distribution peak.
3. I_0/I is the intensity of the central (axial) maximum relative to the maximum intensity within the ring.
4. Δr is the width of the ring at the indicated intensity levels.

The results shown in Fig. 1.23 and Table 1.1 imply that even for binary diffractive elements, the field intensity distribution function can be corrected in the focal area by choosing R_0. Similar effects are achieved also for elements that focus radiation onto an area of arbitrary shape, for instance, onto longitudinal or transversal segments, an ellipse, a transversal (relative to the optical axis) smooth curve, etc. Furthermore, it is not difficult to show that the statements made above are also valid for phase-type diffractive elements.

The problem of synthesizing the structure of the focusing diffractive element thus does not have a unique solution and the field intensity distribution function can be corrected in the focal plane by choosing the free parameter R_0 [65].

1.4.4 Reference Phase in Diffractive Antennas

For the simplest case of the source at infinity, the incident wave fronts are plane waves and the specification of the radii of FZP has a simple form [66]:

$$r_n = \sqrt{n\lambda\sqrt{F^2 + r_0^2} + (n\lambda/2)^2 + r_0^2}\ .$$

The relation between the reference radius and the reference phase is given below.

As shown in Fig. 1.21a, rays from a source (S) pass through a diffractive element to a detection point (F). The rays have a phase at F, which depends on the positions of S and of F, the distance between them, and on the point where the ray went through the diffractive element. The path length of the shortest ray connecting S and F, sometimes called the *direct ray*, is R=|SO|+|OF|. Historically, beginning with Fresnel, the phase of a general ray with path length $r=r_1+r_2$ (where $r_1=|SM|$ and $r_2=|MF|$) has been computed by subtracting R from r, implicitly assuming that the distance R of the shortest ray defines the phase origin $\theta_{ref}=0°$ or reference phase. All rays with phase in the range $-90° < \text{phase} < 90°$ are defined as constructive and a zone plate of the blocking type is designed to block all rays whose phase fall outside this range.

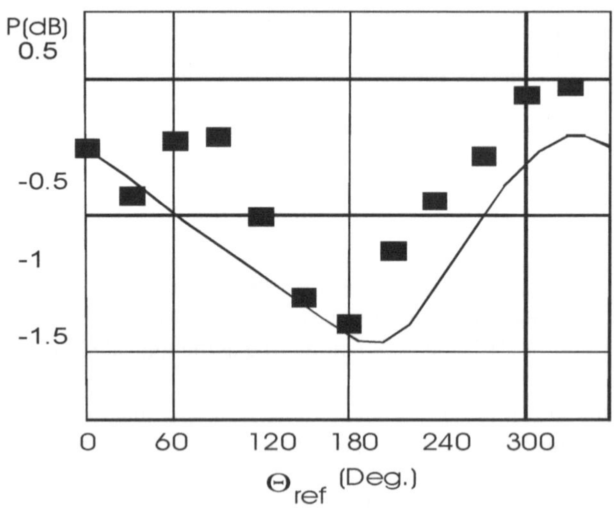

Fig. 1.25 39 GHz beam power, P, as a function of θ_{ref} with 0 dB corresponding to $\theta_{ref}=0°$. Points are measured, solid line is calculated (after [67])

The definition of ray phase can be generalized [67] to include θ_{ref} explicitly:

$$Phase = (r_1 + r_2 - R)\frac{360^\circ}{\lambda} - \Theta_{ref}\,.$$

Although standard zone plate construction assumes a specific choice for reference phase, namely θ_{ref} =0°, it can be chosen to have any value between 0° and 360°. Figure 1.25 displays measured beam power for a set of zone plates with θ_{ref} varied in increments of 30°. Initially, the aperture contained an even number of zones and thus nearly equal areas of in-phase and out-of-phase zones as θ_{ref} is varied through 360°. The solid line is calculated from Fresnel-Kirchhoff. Note that the calculation qualitatively reproduces the magnitude of the overall variation of about 1.7 dB, the existence of a maximum at θ_{ref} =330°, and the minimum near θ_{ref}=180°.

The phase of the focused beam was also measured for the same set of zone plates as in Fig. 1.25. The beam phase data, θ_{beam}, are plotted vs. θ_{ref} as points in Fig. 1.26, relative to that of the zone plate with θ_{ref} =0°. Calculated phase is also shown in Fig. 1.26 by the solid line and is plotted relative to θ_{ref} =0°. Note that measured and calculated phases both vary close to linearly with θ_{ref}. It is evident that the choice of θ_{ref} in the design of a zone plate allows the beam phase to be varied over a complete 360°.

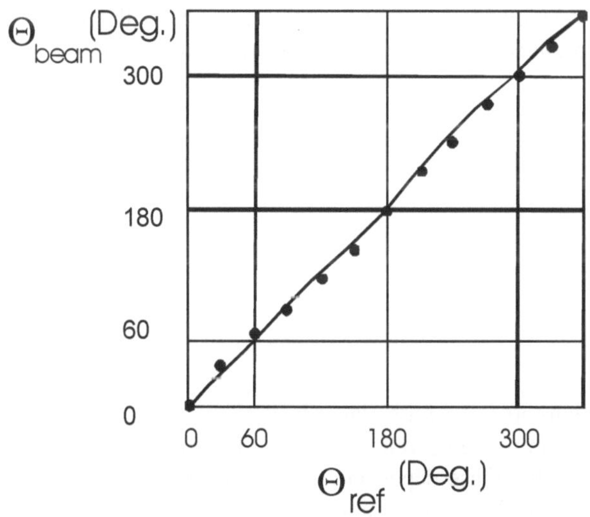

Fig. 1.26 Focused beam phase vs. Θ_{ref} relative to Θ_{ref} =0°. Points are measured data and the solid line is the calculated phase (after [67])

Measured antenna patterns show that the magnitude and position of the maximum side lobe vary with θ_{ref} [67]. The antenna pattern results are summarized in Fig. 1.27, which displays measured data for peak to side lobe ratio (PSL) in dB vs. θ_{ref} in the two cardinal planes. It is seen that the variation of PSL with θ_{ref} tracks approximately for the two directions. It is interesting that the PSL is maximized for the nonstandard value of θ_{ref} =60° and exceeds that of the standard value for a range of θ_{ref}. Figure 1.27 shows that the peak to side lobe ratio can be improved by 2–2.5 dB over the standard construction with θ_{ref} =0°.

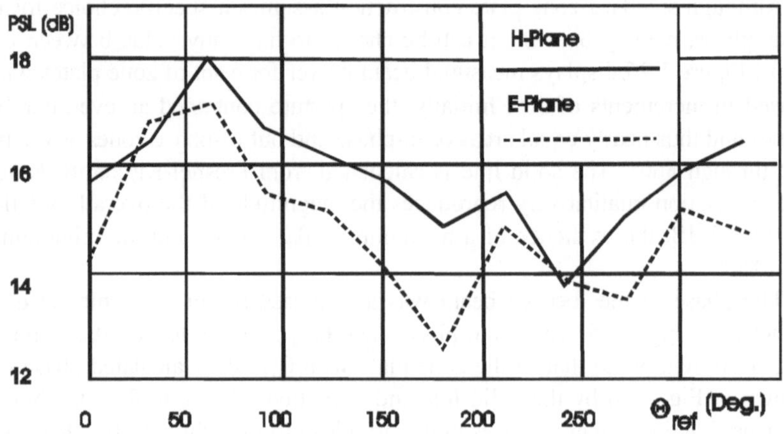

Fig. 1.27 Measured peak to side lobe ratio (PSL) in H-plane and E-plane scans. Note that the optimum PSL occurs for the nonstandard value of θ_{ref} =60° (after [68])

A general relationship between the reference phase, θ_{ref}, and the reference radius, r_0, can be solved for directly (for plane incident wavefront):

$$\theta_{ref} = \frac{360^0}{\lambda}\left(\sqrt{F^2 + r_0^2} - F\right)$$

or

$$r_0 = \sqrt{\left(\frac{\theta_{ref}\lambda}{360^0} + F\right)^2 - F^2}\,.$$

These relations are valid for a planar FZP.

As for the beam phase vs. reference phase for the *curvilinear* FZP, the calculation shows [67] that these dependences are equal to a flat FZP and provides the following relation (with accuracy about 7%):

$$\theta_{beam} \approx \theta_{ref}\,.$$

These results demonstrate that beam phase control by reference radius/phase is a more general new functionality for FZP. For FZP antennas, which can be reconfigured in real time [12, 69, 70, 71], new operational capabilities may be possible.

1.4.5 Reference Phase: Alternative Definition

A recent paper, [65], highlighted the existence of a free parameter in the design of FZPL antennas. This parameter is reference phase (originally introduced in [66]) and, when properly chosen, was shown to improve certain aspects of antenna performance. Though fundamentally the same, reference phase in this work is defined slightly differently than in [65]. This work deals with the specific case of an antenna which, in transmit mode with a source such as a waveguide placed at its focal point, creates a plane wave in the far field. The more general treatment in [65] involves spherical waves from an arbitrary point source on one side of the lens being focused to an arbitrary point on the other side of the lens. The two cases become the same when the arbitrary focus point in the general case is moved to infinity.

Consider again the lens geometry in Fig. 1.28 to specifically indicate the phase. As described previously, the fields from the source travel varying distances to hit the lens surface. The shortest distance traveled is F with all other paths being longer by Δ_i, the path difference. The phase of the shortest path is ϕ_0 and the fields that hit any point on the lens surface will have a radius, r_i, and a phase, ϕ_i. Historically, as it was mentioned above, the radius of the first zone was chosen such that the phase difference, $\phi_1 - \phi_0$, was 180°. This decision was made to simplify calculations and ensure that the out-of-phase waves from 180° to 360° were blocked in order to achieve constructive interference in the far field. However, this phase difference does not have to be 180° and can actually be any value between 0° and 180° (phases between 180° and 360° are actually the dual of the 0°–180° cases, wherein the opaque and transparent regions are simply inversed).

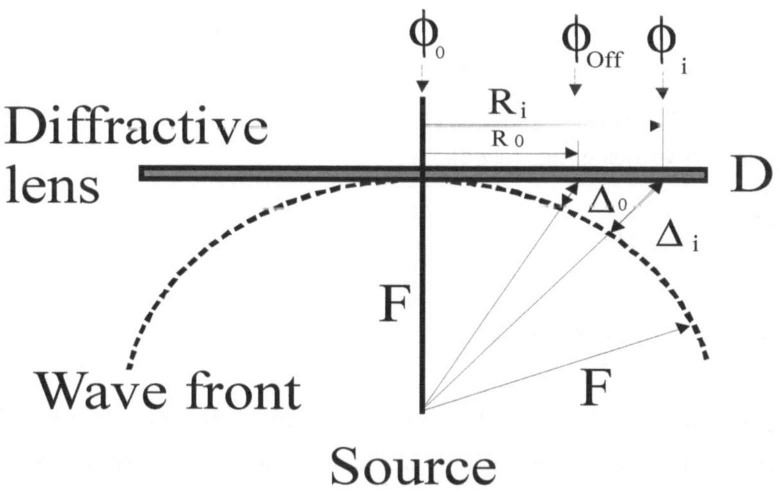

Fig. 1.28 Reference phase definition

Changing the phase difference to between ϕ_1 and ϕ_0 creates a new first zone with radius r_0 and offset phase ϕ_{off}, as illustrated in Fig. 1.28. This causes a shift in the phase of all the other zone radii by the amount of the phase offset since ϕ_1, now the phase of the second zone, must be 180° from ϕ_{off}. The new zone represents a sort of phase correction because instead of allowing the phase difference, or error, to reach 180°, it is corrected at a value less than 180° while still maintaining constructive interference in the far field. Equation 1.29(a) shows how the phase offset is included in the standard phase difference expression with the zone index beginning at zero instead of one in order to include the additional zone.

$$(\phi_i - \phi_0) = i\pi + \phi_{off},\ \text{i=0,1,2,...,N.} \tag{1.29a}$$

$$\text{Let } \Phi_0 = \phi_0 + \phi_{off}. \tag{1.29b}$$

$$\text{Therefore: } (\phi_i - \Phi_0) = i\pi,\ \text{i=0,1,2,...,N.} \tag{1.29c}$$

As a simplifying measure, the sum of ϕ_0 and ϕ_{off} will be represented by Φ_0, which is defined as the reference phase and is expressed mathematically in Eq. 1.29(b). Since ϕ_0 is typically chosen to be 0°, the reference phase becomes equal to the offset phase. Substituting the reference phase into Eq. 1.29(a) yields Eq. 1.29(c).

Antenna designs include this new reference phase parameter in the computation of the Fresnel zone radii. The derivation of this new expression begins with an analysis of the trigonometry in Fig. 1.28:

$$r_i^2 + F^2 = (F + \Delta_i)^2 \tag{1.30}$$

Recognizing that ϕ_i is actually the electrical length of Δ_i, then Eq. 1.14 is simply another version of Eq. 1.29(c).

$$\frac{2\pi\Delta_i}{\lambda} = \phi_i = i\pi + \Phi_0. \tag{1.31}$$

$$\text{Therefore: } \Delta_i = \frac{\lambda}{2\pi}(i\pi + \Phi_0).$$

Substituting Δ_i from Eq. (1.31) into Eq. (1.30) yields the result of Eq. (1.32), which is the final expression for zone radii when reference phase is included. Careful inspection of Eq. (1.32) confirms that increasing the reference phase effectively increases the radii of the Fresnel zones and hence the overall diameter of the lens.

$$r_i = \sqrt{\left(F + \frac{\lambda}{2\pi}(i\pi + \Phi_0)\right)^2 - F^2}\ ,\ i = 0,1,2,..,N \tag{1.32}$$

It can be shown that, if Φ_0=0°, Eq. (1.32) is transformed into classical.

Comparing the expression in Eq. (1.32) with that from [70], it can be seen that Θ_{ref} from [65] is actually ($\Phi_0 + \pi$) in this consideration. This represents a 180° difference between the two above-mentioned reference phases.

1.5 Main Principles of Real-Time Reconfigurable Diffractive Antenna Design

1.5.1 Theory of Photoconductivity FZP

Control of the electronic properties of photoconductors with light has a long history. Historically, silicon or germanium plates that change their properties under illumination by light were used by Dolmatova in the microwave band as converters [12, 68]. This paper suggested creation of semiconductor zone plates by generating annular plasma inhomogeneities in the semiconductor. The zone plate was fabricated from 1.5 mm thick n-type germanium plate, with resistivity of 40 Ω cm, in which plasma was initiated by illumination through a mask or by projecting an image of the zone plate on the surface of the semiconductor.

As a rule, the radii of the zone-plate rings are determined from geometric considerations [12]. For the n-th ring, the radius ρ_n becomes:

$$\rho_n = \sqrt{nf\lambda + \left(\frac{n\lambda}{2}\right)^2}\ . \tag{1.33}$$

where f is the focal length of the zone plate and A is the wavelength in the substrate.

Illuminated areas correspond to regions of high-concentration electron-hole plasma and therefore, to low transparency for millimeter wavelength radiation. Shaded areas in their turn correspond to regions of low concentration of electron-hole plasma, that is, to low transparency. When light source is turned on, a Fresnel zone plate is created in the path of the electromagnetic field, whose zone radii can be varied smoothly if a variable focal distance lens is used. Experiments showed that it is possible to change the conditions of creating nonuniform plasma in a semiconductor and the ratio of conductivities in different areas of the semiconductor, and therefore, control the position of the focal spot, its size, and the intensity of radiation at this point. If the transparency can be made movable along the surface of the semiconductor, a scanning system may be created.

This capability for optical control of electronic properties is under active investigation in a variety of alternative approaches to beam scanning, which employ photoconductors [69, 72, 73, 74, 75, 76].

It is interesting to note that it took nearly twenty years for these ideas to be found useful in developing a number of new devices; for instance, optically controlled millimeter wavelength band modulators [77].

Following [78] let us consider the theory of photoconductivity FZP. A simple analysis was developed for the circular FZP antenna, which is based on the plane wave spectrum for an antenna transmitting into a material half-space [79]. The radiated field of the MMW source is represented by a spectrum of plane waves propagating outward in all directions. The amplitudes of the plane waves are given by the spectral density function, which is derived from the current distribution of the source. Each individual plane wave after interacting with the semiconductor body is calculated using Fresnel reflection and transmission coefficients.

The MMW source is at the origin of the unprimed coordinate system, and the zone plate at the origin of the primed coordinate system. The coordinates of the observation point in air are given in the unprimed system (r, θ, ϕ). The coordinates of a point on the zone plate are also given in an unprimed system, but now denoted by the subscript **p** (r_p, θ_p, ϕ_p).

The MMW source is a rectangular waveguide mounted on an infinite ground plane and with dimensions of *b* and *a* in height and in width, respectively. The aperture field is approximated by the dominant TE_{10}-mode. The expression for the electric field of the waveguide radiating into a material half-space located at a height *h* above the planar interface is given as follows [78]:

$$\vec{E}_1 = \frac{j\exp(-jkr_p)}{2\pi r_p} k_1 \left|\cos\theta_p\right| \left\{ \hat{\theta} \begin{bmatrix} A_{11}^-(\vec{K}_{s1})\exp(jk_1h|\cos\theta_p|) + \\ R_{11}(K_{s1})A_{11}^+(\vec{K}_{s1})\exp(-jk_1h|\cos\theta_p|) \end{bmatrix} + \hat{\phi} \begin{bmatrix} A_\perp^-(\vec{K}_{s1})\exp(jk_1h|\cos\theta_p|) + \\ R_\perp(K_{s1})A_\perp^+(\vec{K}_{s1})\exp(-jk_1h|\cos\theta_p|) \end{bmatrix} \right\} \quad (1.34)$$

with

$$\vec{K}_{si} = k_i \sin\theta[\hat{x}\cos\phi + \hat{y}\sin\phi], \ K_{si} = \kappa, \sin\theta \quad i = 1,2 \quad (1.35)$$

where K_{si} is the transverse component of the plane-wave propagation vector. $R_{II.\perp}(K_{si})$ are the Fresnel reflection coefficients evaluated at K_{si} and depending on the region of observation:

$$R_{II}(K_{s1}) = \frac{k_{21}^2 \,|\cos\theta_p| - \sqrt{k_{21}^2 - \sin^2\theta_p}}{k_{21}^2 \,|\cos\theta_p| + \sqrt{k_{21}^2 - \sin^2\theta_p}},$$

$$R_{\perp}(K_{s1}) = \frac{|\cos\theta_p| - \sqrt{k_{21}^2 - \sin^2\theta_p}}{|\cos\theta_p| + \sqrt{k_{21}^2 - \sin^2\theta_p}}, \tag{1.36}$$

$$k_{21} = \frac{1}{k_{12}} \equiv \frac{k_2}{k_1}.$$

A is the spectral-density and k_1 and k_2 are the complex wave numbers in region 1 and 2, respectively. The components of A are easily computed from the surface current densities and given as follows:

$$A_{II}^{\pm}(\vec{K}) = \mp \frac{\zeta_1}{2K_{si}} \int_{-\infty}^{\infty}\int_{-\infty}^{\infty} \vec{K}\cdot\vec{K}_e \exp(j\vec{K}\cdot\vec{r}')ds',$$

$$A_{\perp}^{\pm}(\vec{K}) = \mp \frac{k_1\zeta_1}{2\gamma_1 K_{si}} \int_{-\infty}^{\infty}\int_{-\infty}^{\infty} \vec{K}\cdot(\hat{z}'\times\vec{K}_e)\exp(j\vec{K}\cdot\vec{r}')ds'. \tag{1.37}$$

K_e is the surface current density vector, K is the plane-wave propagation vector, ζ_1 is the complex wave impedance, and γ_1 is the wave number in the z-direction. Equation (1.36) is calculated for the dominant TE_{10}-mode aperture field with amplitude of E_0 and is given by [80]:

$$A_{II}^{\pm}(\vec{K}) = \mp \frac{\zeta_1 k_x E_0}{2K_1}\left(\frac{\pi a}{2}\right)\left[\frac{b\sin(\frac{b}{2}k_y)}{\frac{b}{2}k_y}\right]\left[\frac{\cos(\frac{a}{2}k_x)}{\left(\frac{a}{2}k_x\right)^2 - \left(\frac{\pi}{2}\right)^2}\right],$$

$$A_{II}^{\pm}(\vec{K}) = \mp \frac{\zeta_1 k_x E_0}{2K_1}\left(\frac{\pi a}{2}\right)\left[\frac{b\sin(\frac{b}{2}k_y)}{\frac{b}{2}k_y}\right]\left[\frac{\cos(\frac{a}{2}k_x)}{\left(\frac{a}{2}k_x\right)^2 - \left(\frac{\pi}{2}\right)^2}\right]$$

$$k_y = k_1\sin\theta_p\cos\phi_p,\; k_x = k_1\sin\theta_p\sin\phi_p,\; \gamma_1 \equiv k_z = k_1\cos\theta_p. \tag{1.38}$$

Substitution of Eqs. (1.38) and (1.34) gives the far-zone electric field of the waveguide radiating into a material half-space. The expression for the field in the substrate (region 1) is used to calculate the physical optic current K_e for the zone-plate rings:

$$\vec{K}_e = 2(\hat{n} \times \vec{H}_1). \tag{1.39}$$

The x` and y` components of this surface current are:

$$K_{e,x'} = \frac{2}{\zeta_1} \cos\theta_p \exp(jk_1 h|\cos\theta_p|) \left\{ \begin{matrix} \cos^2\phi_p[1 - R_{II}\exp(-2jk_1 h|\cos\theta_p|)] - \\ \sin^2\phi_p[1 + R_\perp \exp(-2jk_1 h|\cos\theta_p|)] \end{matrix} \right\} MQ$$

$$K_{e,y'} = -\frac{2}{\zeta_1} \exp(-jk_1 h|\cos\theta_p|)[\cos\theta_p \cos\phi_p \sin\phi_p (R_{II} + R_\perp)]MQ$$

$$Q = \left(\frac{\pi a}{2}\right) \left[\frac{b \sin\left(\frac{b}{2}k_y\right)}{\frac{b}{2}k_y} \right] \left[\frac{\cos\left(\frac{a}{2}k_x\right)}{\left(\frac{a}{2}k_x\right)^2 - \left(\frac{\pi}{2}\right)^2} \right] \tag{1.40}$$

$$M = jE_0 \omega \mu_0 \frac{\exp(-jk_1 r_p)}{4\pi r_p}.$$

μ_0 is the permeability of the free space, ω is the angular frequency.

The spectral density function for the current on the zone-plate rings is determined from Eq. (1.37). The far-zone electric field produced in the air is given as follows:

$$\vec{E}_2 = \frac{j\exp(-jkr)}{2\pi r} \exp(-jk_1 h \sqrt{1 - k_{21}^2 \sin^2\theta}) k_2 |\cos\theta| \left\{ \begin{matrix} \hat{\theta} A_{11}^+(\vec{K}_{s2}) T_{II}(K_{s2}) + \\ \hat{\phi} A_\perp^+(\vec{K}_{s2}) T_\perp(K_{s2}) \end{matrix} \right\} \tag{1.41}$$

$T_{Ut} \pm\{K_{Si})$ are the Fresnel transmission coefficients evaluated at K_{si} depending on the region of observation:

$$T_{II}(K_{s2}) = \frac{2k_{12}\sqrt{k_{12}^2 - \sin^2\theta}}{k_{12}^2 |\cos\theta| + \sqrt{k_{12}^2 - \sin^2\theta}},$$

$$T_\perp(K_{s2}) = \frac{2\sqrt{k_{12}^2 - \sin^2\theta}}{|\cos\theta| + \sqrt{k_{12}^2 - \sin^2\theta}}. \tag{1.42}$$

The variables of integration over the rings are changed from x' and y' into p and ϕ_p, and the following substitutions are made:

$$r_p = \sqrt{f^2 + p^2}, |\cos\theta_p| = \frac{f}{\sqrt{f^2 + p^2}}, \sin\theta_p = \frac{p}{\sqrt{f^2 + p^2}}, \quad (1.43)$$

where p is the radial distance in the plane of the zone plate. The components of the far-zone electric field are given by:

$$E_r^2(r,\theta,\phi=\frac{\pi}{2}) = -\hat{\theta}\frac{j\exp(-jk_2 r)}{2\pi r}\exp\left[-jk_1 f\sqrt{1-k_{21}^2\sin^2\theta}\right]k_2|\cos\theta|T_{II}(K_{s2})$$

$$\sum_{n=1}^{N_{HZ}}(-1)^n \int_{\phi_p=0}^{\phi_p=2\pi}\int_{p_{n-1}}^{p_n}\left[\sin\phi_p \frac{f}{\sqrt{f^2+p^2}}\cos\phi_p\right] \quad (1.44)$$

$$\exp\left[-jk_1\frac{f^2}{\sqrt{f^2+p^2}}\right]\left[R_{II}(K_{s2}) + R_{\perp}(K_{s2})\right]M(p,\phi_p)$$

$$Q(p,\phi_p)\exp(jp\sin\phi_p k_2\sin\theta)\,pdpd\phi_p$$

and

$$E_r^2(r,\theta,\phi=0) = -\hat{\phi}\frac{j\exp(-jk_2 r)}{2\pi r}\exp\left[-jk_1 f\sqrt{1-k_{21}^2\sin^2\theta}\right]k_2|\cos\theta|T_{\perp}(K_{s2})$$

$$\sum_{n=1}^{N_{HZ}}(-1)^n \int_{\phi_p=0}^{\phi_p=2\pi}\int_{p_{n-1}}^{p_n}\frac{f}{\sqrt{f^2+p^2}}\exp\left[jk_1\frac{f^2}{\sqrt{f^2+p^2}}\right] \quad (1.45)$$

$$\left\{\begin{array}{l}\cos^2\phi_p\left[1+R_{II}(K_{s2})\exp\left[-2jk_1\frac{f^2}{\sqrt{f^2+p^2}}\right]\right]+\\ \sin^2\phi_p\left[1+R_{\perp}(K_{s2})\exp\left[-2jk_1\frac{f^2}{\sqrt{f^2+p^2}}\right]\right]\end{array}\right\}$$

where $M(p, \phi_p)$ and $Q(p, \phi_p)$ are given in Eq. (1.40). N_{HZ} is the total number of half zones, where a half zone is either an actual zone-plate ring or a fictitious zone-plate ring, and p_{p-1} and p_n are given by (1.33).

Based on similar principles, a beam-scanning antenna was analyzed and designed at Ka band [81]. The antenna was based on the principle of the Fresnel zone-plate antenna. The scanning of the beam is achieved by using a laser to project different patterns onto a wafer.

1.5.2 Reconfigurable Plasma FZP

The technique described below [69] was based upon a transient Fresnel zone plate approach in which a spatially varying density of charge carriers is created by optical injection of plasma into a semiconductor or photoconductor wafer. The photo-injected charge carriers, electrons, and/or holes alter the index of refraction of the wafer locally, particularly the imaginary part of the index, and thereby attenuate and reflect incident electromagnetic radiation. Conversely, with suitable choice of parameters, the un-illuminated parts of the semiconductor allow the incident electromagnetic waves to be transmitted. The wafer, modified by light in this way, can be made to diffract incident radiation into a beam, which can be scanned. Because the wafer responds rapidly to changes in optical injection, it is possible to rapidly change the Fresnel diffractive conditions and thus rapidly change the beam direction.

For the photoconductor described in [69, 72, 73], a variety of several materials were investigated. Experimental results are given here for Si wafers with dark resistivity in the range of 1000 Ω-cm, diameter of 150 mm, and thickness of 0.7 mm. For photo-injection, light of wavelength below the band gap is used. In early testing, a variable spatial distribution of light intensity was created on the wafer surface by illuminating it through optical masks, which were transmissive to MMW [73, 74, 75]. Since then programmable arrays of individual lights have been employed.

Figure 1.29 shows two configurations [70]. In Figure 1.29a, an MMW feed behind the wafer transmits it through the wafer. A programmable light array projects a light pattern onto the wafer to form the Fresnel lens shaped plasma. In Figure 1.29b, a front feed transmits MMW through the wafer, which is reflected by an MMW reflector that is transparent to the light array. The light array projects a light pattern onto the back of the wafer. In this way, the MMW makes a double pass through the wafer.

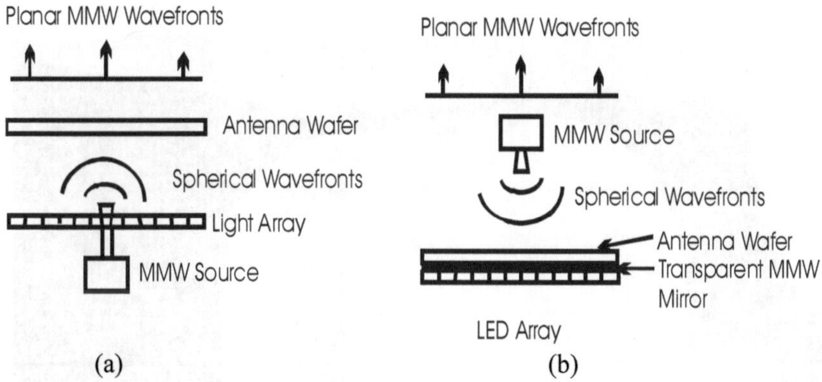

Fig. 1.29 Antenna configurations (**a**) back feed transmissive, (**b**) front feed with transmissive antenna folded around optically transparent MMW mirror (after [69] with permission by Dr. G.W.Webb)

The results with two programmable light arrays (PLA) in the geometry of Fig. 1.26b, one containing 1285 LEDs and the other with 2335 LEDs, with a peak LED output at 0.88 μ are shown [72]. The light array is controlled by a PC through a custom electronic interface. The system can switch the lights on/off at the maximum rate of 140 MHz (140 M lights/sec.) The maximum rate for the 1285 LED frame is 80 K frames per second; however, the overall scan rate with either PLA is limited by the properties of the photoconducting material.

Several light arrays and drive electronics also were developed in [73]. Figure 1.30 [73] displays a PLA consisting of 1285 LEDs having a peak LED output at 0.88 μ. The light array is controlled by a PC through a custom electronic interface, which can switch the lights on/off at the maximum rate of 140 MHz (140 M lights/sec.) The maximum rate is thus 80 K frames per second; however, the overall scan rate is limited by the properties of the photoconducting material to a few kHz.

Fig. 1.30 Programmable light array of 150 mm diameter with 1285 LEDs (after [72] with permission by Dr. G.W.Webb).

Figure 1.31 [73] displays a PLA consisting of 3046 LEDs having a peak LED output at 0.88 μ. The light array electronics is controlled by a PC through a fiber optic interface, which can switch the lights on/off at the maximum rate of 240 MHz (240 M lights/sec.) The maximum rate is thus 80 K frames per second; however, the overall scan rate is limited by the properties of the photoconducting material to a few kHz. An advantage of the fiber optic interface is that it allows the LED electronics package to be made more compact. The overall thickness of the LED array and electronics, neglecting ancillary components, is 6 cm.

A large high-angle side lobe was observed at 94 GHz in some directions at about 55° off the main beam when using the 1285 LED PLA; however, this feature was not observed at lower MMW frequencies. For example, at 94 GHz, the side lobe could be seen at –25° when the beam is scanned to +30° in the H-plane. It was found that the side lobe was an aliasing effect between the finer zone plate features at 94 GHz and the light spacing of the 1285 LED PLA. Accordingly, the light spacing was reduced in the PLA to accommodate a total of 2335 LEDs. This higher light density PLA eliminated the spurious side lobe within the desired scan cone of 30° half angle.

Fig. 1.31 Programmable light array of 350 mm diameter with 3046 infrared emitting LEDs (after [73] with permission by Dr. G.W.Webb)

Thus in [69], the authors demonstrated the ability to form a narrow MMW beam in a solid state scanner by using a photo-injected plasma to block the out-of-phase rays. By controlling which rays need to be blocked, the beam can be rapidly re-formed in different directions. To estimate maximum efficiency or gain, we note that approximately 50% of the rays are blocked compared to the full aperture. Furthermore, the remaining in-phase rays have a distribution of phases over 180° compared, for example, to a fixed parabolic reflector or a phased array with perfect control over phase. From these considerations, we estimate that the maximum gain or efficiency would be down by approximately 10 dB, as presently configured. The advantage, however, is the ability to rapidly steer the beam over a fixed reflector, and at much lower cost than a phased array.

Programmable light sources have been developed, which can produce up to 80 x 10^3 new plasma patterns per second. However, the fundamental limit on scan rate is by the free carrier recombination time in the particular photoconducting material used. Our experiments with this material have shown that we can produce over 1000 independent beams per second, that is, to say by hopping in arbitrary direction with hops of arbitrary magnitude. When the beam is hopped in smaller increments, smaller changes are required in successive plasma distributions. In that case, the effective scan rate increases to above 4000 beams per second.

The similar design of so-called optocontrolled antennas was made in [82]. The optocontrolled antenna system was represented as the aperture covered with a dynamic photosensitive material, whose electromagnetic characteristics depend on the intensity and density of controlling the optical radiation. The projective distribution of controlling optical radiation forms required amplitude-phase distribution for electromagnetic radiation of millimeter-wave band. The amplitude control forms distribution of light intensity as focusing Fresnel zones. The phase control forms the dynamic phase hologram on a multi-layer photosensitive covering. The ways of formation of the dynamic phase hologram on a basis of optical switching elements are investigated. The providing of a highly effective phase method of controlling is complicated because of expensive multi-layer photosensitive covering. The development of a polarizing way allows forming the dynamic phase diagram on single-layer photosensitive covering by means of the optical projective device of controlling.

1.5.3 Electrical and Mechanical Concepts of Reconfigurable Technology [83]

The unique possibility for Fresnel reflectors is to provide solutions to a new antenna marker, where the classical technologies cannot offer good performance. It is known that one of the themes of major interest for the future of telecommunications and remote sensing applications is the study of the antennas with reconfigurable beam pattern capability.

The idea is based on a flat structure formed by small reflectors, whose positions are electronically controlled by means of mechanical actuators, and that works as a reflector with any Fresnel distribution. The lattice of small reflecting cells is electronically controlled to provide any Fresnel distribution over the reflector surface. So, it is possible to have real control of the radiation pattern or/and control of antenna configuration geometry.

The radiation pattern of a reflected signal can be modified, for example, by changes of the geometrical characteristics of the reflecting surface. As it is known, perturbations on parabolic surfaces allow any desired shape of the 3 dB transversal section in the main beam, in the design of reflector antennas in geo-satellite communications.

Looking for an electronic control of the reflecting surface in order to modify the radiation pattern shape of the antenna, the authors [84] present an idea that can be used with many and different applications. This idea (developed by Julio Gutierrez Rios and Juan Vassal'lo Sanz) transforms a classical Fresnel reflector to Active Fresnel Reflector which has fixed the feeder, and then changing the Fresnel zones distribution to have an electronic scanning [84]. Such an Active Fresnel Reflector is formed by a square distribution of reflecting cells, whose position can be modified versus the perpendicular direction to the reflecting surface. Figure 1.32 shows the studied geometry of a distribution of 12 × 12 reflecting cells, in the case of a Fresnel reflector surface of two levels [83].

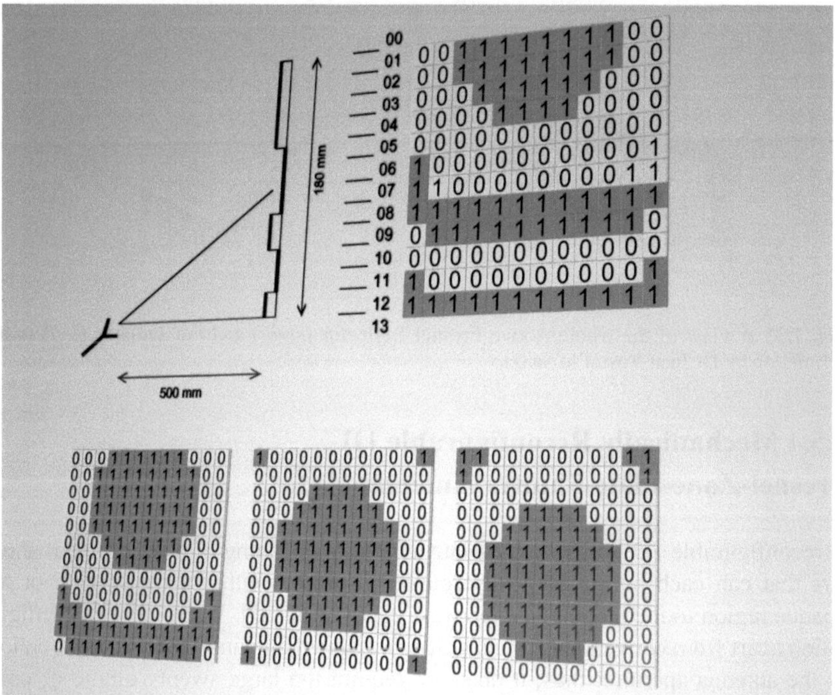

Fig. 1.32 Geometry of a distribution of 12 × 12 reflecting cells (after [84] with permission by Dr.Juan Vassal'lo Sanz)

The selected working frequency was 18.75 GHz, which must be in agreement with the distance separation between both levels (a quarter of the wavelength). The cell size in the antenna described above was determinate taking into account the supposed application of the antenna example, in order to avoid grating lobes, and the area to be covered by the scanning.

The first objective of the mechanical structure is to allow the change of Fresnel level of any reflecting cell. Each Fresnel level is associated with a different position of the reflecting surface of the cell, versus the perpendicular direction to the reflector surface. To provide this position change, the authors [84] selected the commercial actuator that is shown in Fig. 1.33. As it can be seen in Fig. 1.33, the actuator is placed under each reflecting cell, and it allows a displacement of the reflecting cell of 4.5 mm, which is electronically controlled. This figure shows two complete "active reflecting cells", at both states: "0" and "1", which correspond to both Fresnel levels. The actuator is placed at the position "1" when there is a voltage of 6 V at the inputs of the actuator terminal, and it is placed at the position "0" when the voltage is 0. The actuator works over the reflecting cell as it can be seen in Fig. 1.33.

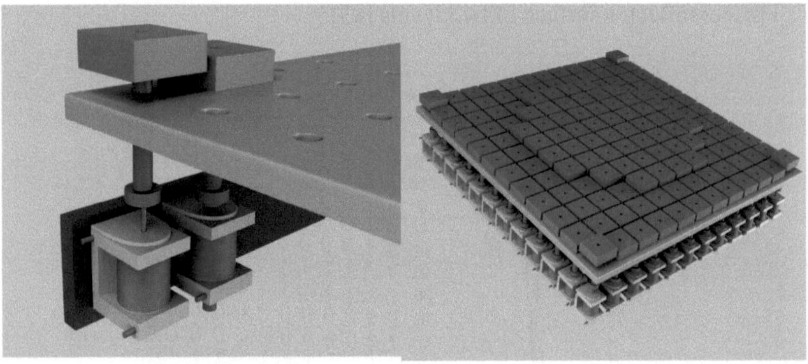

Fig. 1.33 A view of the whole Active Fresnel Reflector (*right*) and the actuator (*left*) (with permission by Dr.Juan Vassal'lo Sanz).

1.5.4 Mechanically Reconfigurable 1D Fresnel-Zone-Plate-Shutter Antenna

A reconfigurable 1D Fresnel-zone-plate antenna, consisting of a set of metal shutters that can each be mechanically rotated to present either a transparent or an opaque region to the incoming wave was proposed in [85]. What sets this configuration apart from conventional mechanically scanned antennas is the fixed location of the antenna aperture: thus, it does not require the large swept volume of conventional mechanically scanned antennas. Only the shutters within the aperture are each rotated to one of two positions. Since the mass of each shutter is relatively small, the scanning rate should be much higher than for conventional mechanically scanned antennas.

In order to give the 1D Fresnel-zone-plate antenna the capability of beam scanning, the aperture is sub-divided into a set of long, thin parallel metallic strips, of width w and thickness t. Each strip can be considered to be a shutter, and can be rotated to an "open" or "closed" position so that it appears either transparent or opaque to an incoming wave (polarized in the x direction). The width of the strip is chosen large enough to appear opaque to the incoming wave in the closed position, while the thickness is chosen small enough to appear nearly transparent in the open position. The selection of the strip width is also a compromise between the discretization resolution and the total number of strips. The smaller the strip width, the better it can approximate various opaque and transparent regions required to scan the beam to the desired angle. However, for a given aperture size, using smaller strips will require a higher total number of strips, which becomes less desirable to implement in a practical design.

Thus, by dividing the antenna aperture into a set of thin metallic shutters that can be individually controlled to create a transparent or opaque sub-zone, various Fresnel-zone patterns can be synthesized to dynamically steer the location of the beam peak. A prototype designed at 23 GHz was fabricated as a proof of concept [85].

1.5.5 2-Bit Antenna–Filter–Antenna Elements

A fully electronic high-resolution scanning requires integration of phase-shifting devices within the array elements to form reconfigurable arrays. A new type of the 2-bit antenna-filter-antenna (AFA) element based on slot antennas and switchable resonators was proposed in [86]. This reconfigurable AFA can operate in four modes of operation as a three- or four-pole filter, and yields a 2-bit variable phase delay. AFAs are three-layer metallic structures composed of receive and transmit antennas and interconnecting resonant circuits, and act as filters with radiation ports [87].

The phase offsets existing between the different modes of operation of the reconfigurable AFA can be utilized to generate a 2-bit adaptive lens array. Elements in the lens array are configured to compensate for the spherical phase delay of the input wavefront and generate a phase distribution corresponding to the desired output wavefront. For a single beam output, the output phase will be a 2-bit approximation of a planar phase distribution. The state of each AFA element in the array can be calculated based on the location of the element in the array, focal distance, output phase, and operation frequency.

Five 32-GHz lens arrays were fabricated in [87] with the AFA elements configured to produce output beams at 0, 15, 30, 45, and 60 from a spherical input wave emanating at a focal distance of 12 cm. By exchanging the roles of input and output sides, a single set of array prototypes can be used to characterize beam steering in both E- and H-planes. 1021 AFA elements are arranged on a rectangular grid to form a circular array with diameter of 12 cm and a total effective area of 10^4 cm. It could be noted as the AFA elements are primarily bandpass filters, the adaptive lens array exhibits a bandpass frequency response. The total loss due to the 2-bit phase quantization error and switches was about 3 dB.

1.5.6 Liquid Crystal Antennas

Generally, the lens antennas are composed of a homogeneous dielectric material or metallic elements. Considering modulation effects, nematic liquid crystal (LC) materials with relatively large dielectric anisotropy in the microwave and millimeter-wave region [88, 89, 83] are attractive for applications to millimeter-wave reconfigurable devices.

The focusing properties of the LC lens [90] and phase shifter [90] at 94 GHz theoretically and experimentally were investigated. In [90], the LC cell is composed of the nematic LC material (E44) with a positive dielectric anisotropy and the planoconvex-shaped metal plates (brass) as electrode substrates. The LC layers and the metal substrates were alternatively piled up. The thickness and radius of the metal substrates were 0.8 and 45 mm, respectively. The thickness of the LC layers was determined by glass ball spacers of 0.3 mm and the maximum cell length was 12 mm. The metal substrates were coated with polyvinyl alcohol (PVA) and treated by rubbing to produce a homogeneous molecular orientation. The polarization direction of the millimeter wave was parallel to the grating vector of the stack-layered structure. The LC cell was driven by 1 kHz sinusoidal ac voltages.

When no external electric field was applied, the director is uniformly oriented parallel to the substrate, and then the effective refractive index of the LC layer is $n_{\perp}$. When a voltage above the threshold was applied across the LC layer, the director is forced to align along the direction of the electric field. Thus, the effective refractive index of the LC layer becomes n_{II}. So in the low voltage regime, the LC molecules in the even zones are re-oriented first while the molecules in the odd zones remain in the original alignment owing to their higher threshold voltage. A binary-phase Fresnel lens is thus formed. If the polarization of the incident wave is parallel to the rubbing direction of the cell, the phase difference between the neighboring zones is

$$\varphi = 2\pi\delta nd/\lambda,$$

where δn is the difference of the effective refractive indices in the adjacent zones, d is the cell gap, and λ is the wavelength of the incident beam. The induced phase shift is electrically tunable. It was shown that the focal length of the LC lens decreases with increasing applied voltage.

The first-order diffraction efficiency of a binary Fresnel lens is determined by the relative phase difference $\Delta\delta$ between the adjacent zones:

$$\Delta\delta = 2\pi d(n_{even} - n_{odd})/\lambda,$$

where d is the cell gap, and n_{even} and n_{odd} are the effective refractive indices of the LC composite in even and odd zones, respectively. The maximum efficiency is reached when $\Delta\delta = \pi$.

In [86], a millimeter-wave beam former using liquid crystal, which can electrically steer millimeter-wave beams by changing the permittivity of the LC was described. It was fabricated by alternately stacking a number of LC layers and electrodes. By employing the membrane impregnated with LC for the LC layer, the structure of the beam former can be simple, avoiding glass substrates and alignment layers. Measurement shows the maximum phase shift to be – 420° at a frequency of 60 GHz, and maximum steering angle to be approximately ±13° and broadband characteristics across the V band. The beam former can also operate as a dielectric lens, which compensates for the quadratic phase variation of the fields over the aperture of the horn antenna. On the other hand, the insertion loss is 5–8 dB and relatively high for practical use. It was found that suppressing the dielectric loss of LC is needed to reduce the insertion loss.

In [91], a novel, nematic liquid crystal cell with a one-dimensional periodic structure prepared using homogeneous and homeotropic orientations alternately as the electrically controlled millimeter-wave devices was proposed. When an external voltage is applied across the LC layer, the LC molecules in the homogeneous orientation tend to align along the applied field direction and the periodicity of the molecular orientation vanishes. The millimeter-wave transmission properties of the LC cell are measured at 50 GHz. Changes in the phase and transmittance of the LC cell induced by applying the voltage to the cell were observed. For frequencies 2–40 GHz, the LC types used were BDH-K15, BDH-E7, and BDH-E44. The maximum phase shift 50°/cm at 37 GHz was observed for a commercial liquid crystal BDH-E44.

Similar investigations also were made in THz waveband [92, 93, 94]. The complex indices of refraction of a nematic liquid crystal, 5CB, PCH5, and E7 were determined to be from 0.2 to beyond 1 THz for the first time. Significantly, the birefringence of 5CB and E7 are as large as 0.2 at 1 THz frequencies, while the absorption is negligible. Electrical-field and magnetic-field-controlled birefringences in LC were also reported. A tunable room-temperature THz phase shifter using magnetic-field-controlled birefringence in nematic 5CB gives a phase shift as large as 108° at 1.0 THz. Phase shift exceeding 360° at 1 THz, an important milestone, was realized by using a sandwiched LC (E7, Merck) cell as thick as 3 mm. The principle and performance of LC-based Lyot filters as phase shifters, modulators, and other photonic devices in THz were described in [95, 96]. The transmitted peak frequency and the bandwidth of the filter are 0.465 THz and 0.10 THz, respectively. The THz wave is separated into an o-ray and an e-ray after passing the first element. These two waves are further separated into an o-o-ray, e-o-ray, o-e-ray, and e-e-ray, respectively, again after passing through the second element. The transmission spectrum manifests the interference among the four peaks of the THz signal.

References

1. G. J. K. Moernaut, D. Orban. The Basics of Antenna Arrays: http://www.orbanmicrowave.com/The_Basics_Of_Patch_Antennas.pdf.
2. C. B. Dietrich. Adaptive Arrays and Diversity Antenna Configurations for Handheld Wireless Communication Terminals, Chapter 3: Antenna Arrays and Beamforming, PhD Thesis, Blacksburg, Virginia, 2000, available at: http://scholar.lib.vt.edu/theses/available/etd-04262000–15330030/unrestricted.

3. W. L. Stutzman, G. A. Thiele, Antenna Theory and Design, John Wiley & Sons, New York, 1981.
4. A. Ishide, R. T. Compton Jr., On Grating Nulls in Adaptive Arrays. IEEE Transactions on Antennas and Propagation, Vol. AP-28, No. 4, July 1980 pp. 467–475.
5. J. Fuhl, D. J. Cichon, E. Bonek. Optimum Antenna Topologies and Adaptation Strategies for SDMA. IEEE GLOBECOM 96. Vol. 1, 1996, pp. 575–580.
6. K. Dietze, C. Dietrich, W. Stutzman. Vector Multipath Propagation Simulator (VMPS), Draft Report, Virginia Tech Antenna Group, April 7, 1999.
7. E. Brookner (Ed.). Practical Phased Array Antenna Systems. Artech House, Boston, MA, 1991.
8. R. J. Mailloux. Phased Array Antenna Handbook. Artech House, Boston, MA, 1994.
9. A. Fresnel. Calcul de l'Intensite de la Lumiere au Centre de l'Ombre d'un Ecran et d'une Ouverture Circulaires Eclairee par une Point Radieux. Oeuvres d'Augustin Fresnel, Vol. 1, Note 1, pp. 365–372 (1866). Reprinted in J. Ojeda-Castanada and C. Gomez-Reino, Selected Papers on Zone Plates, SPIE Milestone Series Vol. MS 128 (1996).
10. R. Guenther. Modern Optics. John Wiley & Sons Inc., New York, 1990.
11 H. D. Hristov. Fresnel Zones in Wireless Links, Zone Plate Lenses and Antennas. Artech House Inc., Boston, MA, 2000.
12. O. V. Minin, I. V. Minin. Diffraction Optics of Millimetre Waves, Institute of Physics Publishing, Bristol, UK, 2004.
13. Y. J. Guo, S.K. Barton. Fresnel Zone Antennas, Kluwer Academic Publishers, Boston, 2002.
14. J. C. Wiltse. History and Evolution of Fresnel Zone Plate Antennas for Microwaves and Millimeter Waves. IEEE Antennas and Propagation Int, Symp. 1999, Special Section, Status and Future of Fresnel Zone Plate Antennas. Orlando FL, July 11–16, 1999, Symposium Digest Vol. 2, pp. 722–725.
15. F. Sobel, F. L. Wentworth, J. C., Wiltse. Quasi-Optical Surface Waveguide and Other Components for the 100- to 300-Gc Region. IRE Transactions on Microwave Theory and Techniques, pp.512–518, November 1961.
16. D. N. Black, J. C. Wiltse. Millimeter-Wave Characteristics of Phase-Correcting Fresnel Zone Plates. IEEE Transactions on Microwave Theory and Techniques, Vol. 35, No. 12, pp. 1122–1129, December 1987.
17. J. E. Garrett, J. C. Wiltse. Antenna Pattern Characteristics of Phase-Correcting Fresnel Zone Plates. IEEE Antennas and Propagation Symposium 1990, pp.1906–1909.
18. C. A. Barrett, J. C. Wiltse. Design Parameters for Zone Plate Antennas. IEEE Antennas and Propagation Symposium, 1992, pp.608–611.
19. A. Petosa, A. Ittipiboon. Fresnel Lens Antenna for Microwave Communication Applications. CRC Report No. 2004–003, Ottawa, Nov. 2004.
20. J. C. Wiltse. Recent Developments in Fresnel Zone Plate Antennas at Microwave/Millimeter Wave. SPIE Conference, pp.146–154, July 1998.
21. A. Petosa, A. Ittipiboon. Shadow Blockage Effects on the Aperture Efficiency of Dielectric Fresnel Lenses. IEE Proceedings Microwave Antennas and Propagation, Vol. 147, No. 6, pp. 451–454, December 2000.
22. I. V. Minin, O. V. Minin, Reduction of the Zone Shadowing Effect in Diffractive Optical Elements on Curvilinear Surfaces. Optoelectronics, Instrumentation and Data Processing, Vol. 40, No. 3, pp. 117–127, 2004.
23. I. V. Minin, O. V. Minin, Diffractive quasioptics. – SibAgs Publisher, 1999, 308p. (in Russian).
24. A. Petosa, N. Gagnon, A. Ittipiboon. Effects of Fresnel Lens Thickness on Aperture Efficiency. ANTEM 2004, Ottawa Canada, pp. 175–178, July 20–23 2004.
25. A. Petosa, N. Gagnon, A. Ittipiboon. Optimization of Dielectric Fresnel Lens Thickness for Maximizing Gain. Submitted to ANTEM 2006, Montreal Canada.
26. D. R. Reid, G. S. Smith. A Full Electromagnetic Analysis of Grooved-Dielectric Fresnel Zone Plate Antennas for Microwave and Millimeter-Wave Applications. IEEE Transactions on Antennas and Propagation, Vol. 55, No. 8, August 2007, pp. 2138–2146.

27. S. M. StoutGrandy, A. Petosa, I. V. Minin, O. V. Minin, J. S. Wight. Novel Reflector Backed Fresnel Zone Plate Antenna. Microwave and Optical Technology Letters Vol. 49, No. 12, December 2007, pp. 3096–3098.
28 S. Stout-Grandy, A. Petosa, I.V. Minin, O.V. Minim, J.S. Wight, "Investigation of Low-Profile Fresnel Zone Plate Antennas", URSI North American Radio Science Meeting, Ottawa Ontario, July 23, 2007.
29. R. Merlin. Radiationless Electromagnetic Interference: Evanescent-Field Lenses and Perfect Focusing. Science, Vol. 317, August 2007, pp. 927–929.
30. A. Petosa, A. Ittipoon. A Fresnel Lens Designed Using a Perforated Dielectric. ANTEM 2002, St-Hubert, Quebec, Canada, pp. 399–402, July 31-August 2, 2002.
31. A. Petosa, A. Ittipoon. Design and Performance of a Perforated Dielectric Fresnel Lens. IEE Proceedings Microwave Antennas Propagation, Vol. 150, No. 5, October 2003, pp. 309–314.
32. M. J. Vaughan, K. Y. Hur, R. C. Compton. Improvement of Microstrip Patch Antenna Radiation Patterns. IEEE Transactions on Antennas and Propagation, Vol. 42, No. 6, June 1994, pp. 882–885.
33. G. O. Gauthier, A. Courtay, G. M. Rebeiz. Microstrip Antennas on Synthesized Low Dielectric-Constant Substrates. IEEE Transactions on Antennas and Propagation, Vol. 45, No. 8, August 1997, pp. 1310–1314.
34. J. B. Muldavin, G. M. Rebeiz. Millimeter-Wave Tapered-Slot Antennas on Synthesized Low Permittivity Substrates. IEEE Transactions on Antennas and Propagation, Vol. 47, No. 8, August 1999, pp. 1276–1280.
35. J. S. Colburn, Y. Rahmat-Samii. Patch Antennas on Externally Perforated High Dielectric Constant Substrates. IEEE Transactions on Antennas and Propagation, Vol. 47, No. 12, December 1999, pp. 1785–1794.
36. P. F. Goldsmith. Perforated Plate Lens for Millimeter Quasi-Optical Systems. IEEE Transactions on Antennas and Propagation, Vol. 39, No. 6, June 1991, pp. 834–838.
37. A. Petosa, A. Ittipoon. A Fresnel Lens Designed Using a Perforated Dielectric. ANTEM 2002, St-Hubert, Quebec, Canada, pp. 399–402, July August 2, 2002.
38. A. Petosa, A. Ittipoon. Design and Performance of a Perforated Dielectric Fresnel Lens. IEE Proceedings Microwave Antennas Propagation, Vol. 150, No. 5, October 2003, pp. 309–314.
39. I. Kadri, M. Britton, L. Roy. A Multi Frequency Fresnel Lens Using a Perforated Dielectric. IEEE CCEC-CCGEI 2004, Niagara Falls, pp. 913–916, May 2004.
40. Y. J. Guo, S. K. Barton. Fresnel Zone Plate Reflector Incorporating Rings. IEEE Microwave and Guided Wave Letters, Vol. 3, No. 11, November 1993, pp. 417–419.
41. Y. J. Guo, S. K. Barton. Phase Correcting Zonal Reflector Incorporating Rings. IEEE Transactions on Antennas and Propagation, Vol. 43, No. 4, April 1995, pp. 350–355.
42. B. D. Nguyen, C. Migliaccio, C. Pichot. 94 GHz zonal ring reflector for helicopter collision avoidance. Electronics Letters, Vol. 40, No. 20, September, 2004.
43. B. D. Nguyen, C. Migliaccio, Ch. Pichot, K. Yamamoto, N. Yonemoto. W-Band Fresnel Zone Plate Reflector for Helicopter Collision Avoidance Radar. IEEE Transactions on Antennas and Propagation, Vol. 55, No. 5, May 2007, pp. 1452–1456.
44. M. S. Mirotznik, D. W. Prather, J. N. Mait et al. Three-Dimensional Analysis of subwavelength Diffractive Optical Elements with the Finite-Difference Time-Domain Method. Applied Optics Vol. 39, 2000, pp. 2871–2880.
45. D. Feng, Y. Yan, G. Jin, S. Fan. Beam Focusing Characteristics of Diffractive Lenses with Binary Subwavelength Structures. Optics Communications Vol. 239, 2004, pp. 345–352.
46. M. Born, E. Wolf. Principles of Optics, 7th ed., Cambridge University Press, 1999.
47. H. A. Mcleod, Thin Film Optical Filters, 2nd ed., Mcgraw Hill, New York, 1989.
48. P. Yeh. Electromagnetic Propagation in Birefringent Layered Media. Journal of the Optical Society of America, Vol. 69, 1979, pp. 742–756.
49. M. S. Mirotznik, D. M. Pustai, D. W. Prather, J. N. Mait. Design of Two-Dimensional Polarization-Selective Diffractive Optical Elements With Form-Birefringent Microstructures, Applied Optics Vol. 43, 2004, pp. 5947–5954.

50. M. S. Mirotznik, T. Creazzo, S. Mathews. Design of Diffractive Elements at Millimeter Wavelengths Using Subwavelength Cylindrical Microstructures. Microwave and Optical Technology Letters, Vol. 49, No. 8, August 2007, pp. 1880–1884.
51. J. D. Joannopoulos, R. D. Meade, J. N. Winn. Photonic Crystals: Molding the Flow of Light. Princeton Univ. Press, Princeton, New Jersey, 1995.
52. K. Sakoda. Optical Properties of Photonic Crystals. 2nd ed. Springer-Verlag, New York, 2005.
53. B. Goss Levi. Progress Made in Near-field Imaging with Light from a Sharp Tip. Physics Today Vol. 52, 1999, 9. 18.
54. E. Flück, N. F. van Hulst, W. L. Vos, L. Kuipers. Near-field Optical Investigation of Three-Dimensional Photonic Crystals. Physical Review E Vol. 68, 2003, p. 015601.
55. G. D. Maluzhinetz, Diffraction Near the Optical Axis of a Zone Plate. Doklady AN SSSR. Vol. 54, No 5. 1946, pp. 403–406.
56. J. L. Soret. Ueber die durch Kreisgitter erzeugten Diffractionsphanomene. Annalen der Physic und Chemie, Vol. 156, 1875, pp. 99–113. Reprinted in J. Ojeda-Castanada and C. Gomez-Reino, ibid.
57. L. Rayleigh, Reprinted in Lord Rayleigh Scientific Papers, Vol. 3, 1887–1892, pp. 74–79. R. W. Wood. Phase Reversal Zone Plates and Diffraction Telescope. Phil. Mag. Series 5, Vol. 45, 1898, pp. 511.
58. I. V. Minin and O. V. Minin. Fesnel Zone Plate Lens and Antennas for Millimeter Waves: History and Evolutions of Developments and Applications. Proceedings of the 25th Int. Conference on Infrared and Millimeter Waves, September 12–15, 2000, Beijing, China, pp. 409–410.
59. A. Engel, J. Steffen, G. Herziger. Laser machining with modulated zone plates. Applied Optics Vol. 13, No 12, 1947, pp. 269–273.
60. A. Fedotowsky, K. Lehovec, Far Field Diffraction Patterns of Circular Gratings. Applied Optics, Vol. 13, No 11. 1947, pp. 2638–2642.
61. A. Fedotowsky, K. Lehovec, Optimal Filter Design for Annular Imaging. Applied Optics Vol. 13, No 12. 1947, pp. 2919–2923.
62. V. P. Koronkevich, I. G. Palchikova, A. G. Poleschuk, Yu. I. Yurlov, Kinoform Optical Elements with Annular Momentum Response: Preprint No 256 IAiE SO AN SSSR. Novosibirsk, 1985. p. 19.
63. N. L. Kazansky. Numerical Experimental Study of Diffraction Characteristics of a Focusator Onto A Ring. Kompyuternaya optika, No 10–11, 1992, pp. 128–144.
64. M. A. Golub, N. L. Kazansky, I. N. Sisakyan, V. A. Soifer. Numerical Experiment with Flat Optics Elements. Avtometria No 1. 1998, pp. 70–82.
65. I. V. Minin, O. V. Minin, Control of Focusing Properties of Diffractive Elements. Kvantovaya elektronika Vol. 17, No 2. 1990, pp. 249–251.
66. Inventor's Certificate (USSR) No 1617398, G 02 B 27/42, I. V. Minin, O. V. Minin, Zone plate, 1988.
67. I. V. Minin, O. V. Minin, G. W. Webb. "Flat and Conformal Zone Plate Antennas with New Capabilities. Proc. of the International Conference on Applied Electromagnetics and Communications – ICECOM 2005, Dubrovnik, Croatia, October 11–14, pp. 405–408.
68. E. A. Dolmatova, Semiconductor Fresnel Zone Plate. Vestn. Har'kovskogo Universiteta. No. 7, 163. 1978, pp. 71–72 (in Russian).
69. G. W. Webb, S. C. Rose, M. S. Sanchez, J. M. Osterwalder. Experiments on an Optically Controlled 2-D Scanning Antenna. Proc. 1998 Antenna Applications Symposium, Allerton Park, Monticello, IL, September 16–18, 1998 (http://www.ecs.umass.edu/ece/allerton/papers1998/).
70. M. Hajian, G. A. de Vree, L. P. Ligthatin. Electromagnetic Analysis of Beam-Scanning Antenna at Millimeter-Wave Band Based on Photoconductivity Using Fresnel-Zone-Plate Technique. IEEE Antennas and Propagation Magazine. Vol. 45, No. 5, October 2003, pp. 13–25.

71. C.-C. Cheng, A. Abbaspour-Tamijani. Study of 2-bit Antenna-Filter-Antenna Elements for Reconfigurable Millimeter-Wave Lens Arrays. IEEE Transactions on Microwave Theory and Techniques, Vol. 54, No. 12, December 2006, pp. 4498–4506.
72. G. W. Webb, W. Vernon, M. S. Sanchez, S. C. Rose, S. Angello. Optically Controlled Millimeter Wave Antenna. Proceedings International Topical Meeting on Microwave Photonics, Melbourne, Australia, November 17–19, 1999, p. 275.
73. G. W. Webb, W. Vernon, M. S. Sanchez, S. C. Rose, S. Angello. Novel Photonically Controlled Antenna for MMW Communications. Proceedings International Topical Meeting on Microwave Photonics, Melbourne, Australia, 2000, p. 176.
74. G. W. Webb and L. H. Pinck. Light-Controlled MMW Beam Scanner. Proc. 1993 SBMO International Microwave Conference, Vol. II, Sao Paolo, Brazil, IEEE Cat. No. 93TH0555-3, p. 417.
75. G. W. Webb, L. H. Pinck. MMW Beam Scanner Controlled by Light. Proc. Workshop on Millimeter-Wave Power Generation and Beam Control, Huntsville, AL, Special Report RD-AS-94-4, U.S. Army Missile Command, p. 333.
76. A. R. Rosen, P. J. Stabile, R. Amantea, W. J. Janton, D. B. Gilbert, J. K. Butler. Active Millimeter Wave Antenna Array Steered by Optically Induced Gratings. Proc. Workshop on Millimeter-Wave Power Generation and Beam Control, Huntsville, AL, Special Report RD-AS-94-4, U.S. Army Missile Command, p. 341.
77. A. Tateishi, K. Kikuchi. Application of Fresnel Zone Plate to Millimeter Wave Modulator. Microwave and Millimeter Wave Technology Proceedings, 1998. ICMMT '98. 1998 International Conference on, 19.
78. M. Hajian, W. Keizer, B. J. Reits, L.P. Ligthart. Concept of a Scanning Beam Antenna at 35 GHz based on Photoconductivity Technology. Proc. of the 20th ESTEC Antenna Workshop, 18–20 June 1997, The Netherlands, pp. 55–65.
79. G. S. Smith. Directive Properties of Antennas for Transmission into a Material Half-Space. IEEE Transaction on Antennas Propagation, Vol. AP-32, March 1984 pp. 232–246.
80. C. A. Balanis. Antenna theory, analysis and design. 2nd ed. John Wiely & Sons, New York, 1996, (sub Section 12.5.3).
81. M. Hajian, G. A. de Vree, L. P. Ligthati. Electromagnetic Analysis of Beam-Scanning Antenna at Millimeter-Wave Band Based on Photoconductivity Using Fresnel-Zone-Plate Technique // IEEE Antennas and Propagation Magazine. Vol. 45, No. 5, October 2003, pp.13–25.
82. V. I. Karpenko, O. V. Karpenko. Princoples of Designing of Optocontrolled Antennas for Wideband Radiolocation System. Proc. 2nd Int. Workshop on Ultrawideband and Ultrashort Impulse Signals, September 19–22, Sevastopol, Ukraine, 2004, pp. 59–60.
83. M. Tanaka, S. Sato. Millimeter-Wave Detection Properties of Liquid Crystal Prism Cells with Stack-Layered Structure. Japan Journal of Applied Physics Vol. 40, 2001, pp. 4131.
84. J/ Gutiérrez Ríos, J. Vassal'lo Sanz. New Fresnel Reflector Technology with Reconfigurable Beam Pattern. Proc. of the 29th ESA Antenna Workshop on Multiple Beams and Reconfigurable Antennas, Noordwijk, The Netherlands, 18–20 April 2007.
85. A. Petosa, S. Thirakoune, A. Ittipiboon. Reconfigurable Fresnel-Zone-Plate-Shutter Antenna with Beam-Steering Capability. IEEE Antennas and Propagation Magazine, Vol. 49, No. 5, October 2007, pp. 42–51.
86. K. Hirokazu, K. Takao, F. Hideo, N. Toshihiro. Millimeter-wave Beam Former Using Liquid Crystal. IEICE Technical Report. Microwaves, Vol. 103, No. 534, 2003, pp. 71–76.
87. C. C. Cheng, A. Abbaspour-Tamijani, C. Bircher. Millimeterwave Beam-Steering Using an Array of Reconfigurable Antenna–Filter–Antenna Elements.in IEEE MTT-S International Microwave Symposium Digit. 2006, pp. 449–452.
88. F. Guerin, J. M. Chappe, P. Joffre, D. Dolfi. Modeling, Synthesis and Characterization of Millimeter Wave Multilayer Microstrip Liquid Crystal Shifter. Japan Journal of Applied Physics Vol. 36, 1997, p. 4409.
89. T. Kuki, H. Fujikake, T. Nomoto, Y. Utsumi: Trans. IEICE Electron. J84-C (2001) 90 [in Japanese].

90. M. Tanaka, S. Sato. Electrically Controlled Millimeter-Wave Focusing Properties of Liquid Crystal Lens. Japan Journal of Applied Physics, Vol. 41, 2002, pp. 5332–5333.
91. M. Tanaka, S. Sato. Millimeter-Wave Propagation Properties of the Nematic Liquid Crystal Cell with a 1-D Periodic Structure Induced by Different Molecular Orientations. Molecular Crystals and Liquid Crystals, Vol. 434, 2005, pp. 107–112.
92. R. -P. Pan, T. -R. Tsai, C. Wang, C. -Y. Chen, C. -L. Pan, The Refractive Indices of Nematic Liquid Crystal 4, 4-*n*-pentylcyanobiphenyl in the THz Frequency Range, Molecular Crystaland Liquid Crystal., Vol. 409, 2004, pp. 137–144.
93. T. -R. Tsai, C. -Y. Chen, C. -L. Pan, R. -P. Pan, X. -C. Zhang, Room Temperature Electrically Controlled Terahertz Phase Shifter, IEEE Microwave and Wireless Components Lett., Vol. 14, No. 2, February 2004 pp. 77–79.
94. C. -Y. Chen, T. -R. Tsai, C. -L. Pan, R. -P. Pan. Terahertz Phase Shifter with Nematic Liquid Crystal in a Magnetic Field, Applied Physics Letters Vol. 83, No. 22, December 2003, pp. 4497–4499.
95. C. -L. Pan, R.-P. Pan. Liquid-Crystal-Based Devices Manipulate Terahertz-Frequency Radiation. SPIE Newsroom, 10.1117/2.1200702.0676.
96. C. -L. Pan, R. -P. Pan. Recent Progress in Liquid Crystal THz Optics, Proc. SPIE 6135, pp. D1–13, 2006.
97. J. Ojeda-Castanada and C. Gomez-Reino, Selected Papers on Zone Plates, SPIE Milestone Series Vol. MS 128 (1996).
98. C. C. Cheng, A. Abbaspour-Tamijani. Study of 2-bit Antenna–Filter–Antenna Elements for Reconfigurable Millimeter-Wave Lens Arrays. IEEE Transactions on Microwave Theory and Techniques, Vol. 54, No. 12, December 2006, pp. 4498–4506.
99. I. V. Minin, O. V. Minin, Y. R. Triandaphilov, V. V. Kotlyar. Subwavelength Diffractive Photonic Crystal Lens. Progress in Electromagnetic Research B (PIER B) Vol. 7, 2008, pp. 257–264.

Chapter 2
Lens Candidates to Antenna Array

2.1 Circular Fresnel Zone Plate Antennas with Varying Reference Phase

Let us consider the antenna designs take the new reference phase parameter into account in the computation of the Fresnel zone radii [1]. The Fresnel zone radii were computed using Eq. (1.32). The lens parameters for each *F/D* simulation are shown in Table 2.1. Each one was analyzed with the following nine different reference phases: Φ_0=0°, 30°, 45°, 60°, 75°, 90°, 120°, 150°, 180°. The number of zones was kept constant with increasing reference phase, which meant that the radii of the zones were increasing slightly.

Table 2.1 Simulation parameters for each *F/D*

Nominal *F/D*	$F(\lambda)$	$D(\lambda)$ Desired	$D(\lambda)$ Actual	Number of total zones
0.237	3.75	15.8	10.05	10
0.35	5.53	15.8	15.49	8
0.5	7.9	15.8	15.02	6
1.0	15.8	15.8	16.39	4

Since it was desirable to have a consistent aperture size for comparison purposes, the lens diameters were chosen to be as close to 15.8 cm as possible for the Φ_0=0° case, while still maintaining an integer number of zones. This meant that careful consideration of the number of zones was required for each *F/D* to match the desired diameter of the last zone. Inevitably, the actual lens diameters were not exactly 15.8 cm and are shown in Table 2.1 for the Φ_0 = 0° case. Note that since the radii increased as the reference phase increased, the *F/D* inherently decreased by a small amount from 0° to 180°. This was found to have very little effect on the antenna performance.

The lenses were fed with an open-ended WR28 waveguide placed at their focal point. The waveguide was oriented such that the electric field vector was along the y-axis. The feed shown to have a peak directivity of about 7 dB, with 10-dB beamwidths of 120° in the H-plane and 194° in the E-plane. Despite its inherently unequal H- and E-plane patterns, this feed yields about a 10-dB edge taper for an *F/D* of 0.237. The 10-dB edge taper is commonly used as a compromise between spillover and taper illumination for the parabolic reflector and is also a good approximation for the FZPL.

Figure 2.1 illustrates how a typical co-polarized radiation pattern in the H-plane varies with reference phase.

This is the *F/D* = 0.237 case, and only the first 20° from boresight are shown in order to magnify the effect of the first sidelobe. As the reference phase increased from $\Phi_0 = 0°$ up to 180°, the radiation pattern changed shape and dropped in level.

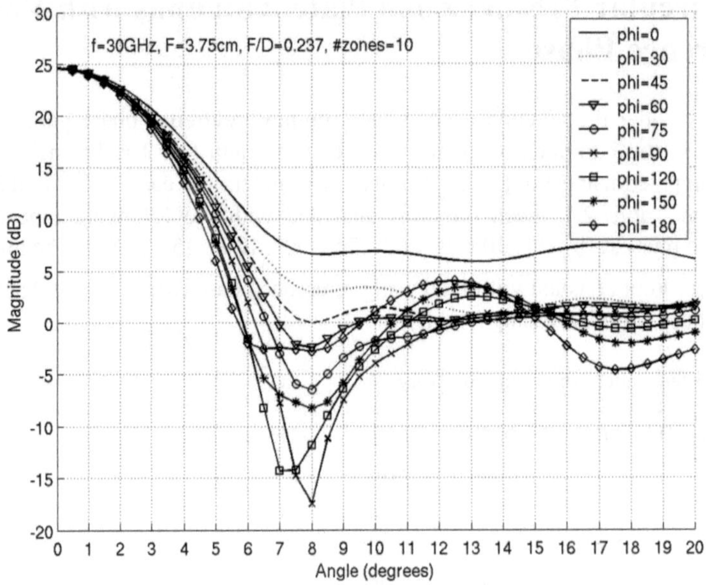

Fig. 2.1 Typical radiation pattern in the H-plane varies with reference phase

The detailed analyses on the effect of reference phase on the different antenna configurations shown the following [1]:

1) *Peak Directivity:* The simulated peak directivity was confirmed to occur at the designed center frequency of 30 GHz for the smaller *F/D* cases (0.237, 0.35). However, this was not true for the larger *F/D* cases where the frequency of the peak directivity was observed to shift closer to 31 GHz. This increase in frequency is likely related to the feed not being optimal for the lens in the larger *F/D* cases. The feed was designed to be optimal for the *F/D* = 0.237 case so when the *F/D* was increased, the feed's 10 dB beamwidth was much larger than that required for the 10-dB edge taper and therefore significant spillover resulted.

The normalized peak directivity for each *F/D* case is shown in Fig. 2.2. The curves were normalized to the peak of the $\Phi_0 = 0°$ case to compensate for the feed not being optimal in the larger *F/D* cases. The variation of the peak directivity with reference phase is related to the fact that changing the reference phase alters the diffraction pattern of the lens in a complex fashion. This caused the sidelobes to shift and change in level, thus impacting the peak gain.

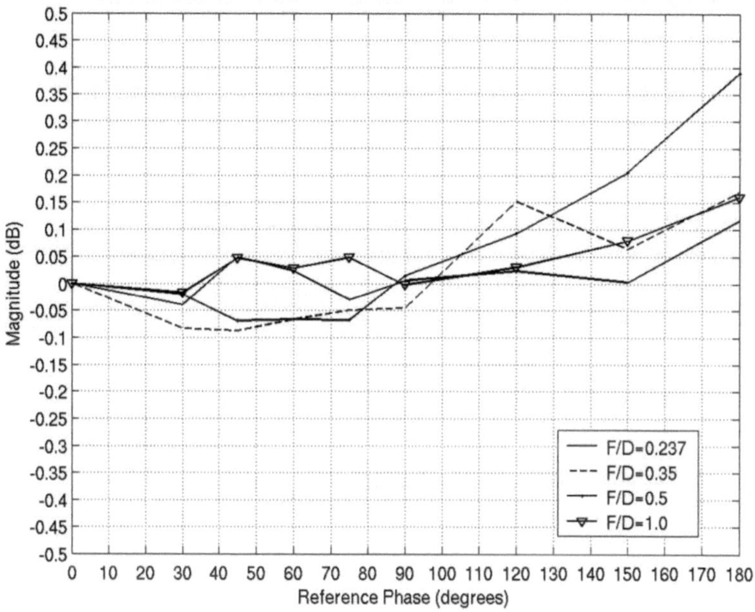

Fig. 2.2 Normalized peak directivity vs. reference phase

The curves of Fig. 2.2 demonstrate that the average change in peak directivity was 0.23 dB as the reference phase increased from 0° to 180°. This means that the reference phase had very little effect on the peak directivity. The gain was also computed from these results based on the simulated losses in the structure, which include material loss, mismatch loss at the excitation port, and loss due to surface resistance. It was found that there was a 2.45 dB difference between the directivity and gain for the *F/D* = 0.237 case. This difference dropped to 1.23 dB for the *F/D* = 1.0 case.

2) 1st Sidelobe Level: The level of the first sidelobe was affected significantly by the change in reference phase. Figure 2.2 illustrates these simulated results for the H- and E-planes, respectively, where the curves in both figures are relative to the peak directivity. The sidelobes were generally better for the smaller *F/D* values. This was true particularly when the reference phase is 0°.

In both the H- and E-planes, all *F/D* cases exhibited a similar trend in that the level of the first sidelobe decreased as the reference phase increased from 0° through to about 60°. When the reference phase increased beyond 60°, the sidelobe also increased such that at 180° the sidelobe was nearly the same as it was at 0°. The overall decrease in sidelobe, as well as the normalized minimum sidelobe and the reference phase where this minimum occurred, are shown in Table 2.2 for each *F/D* case.

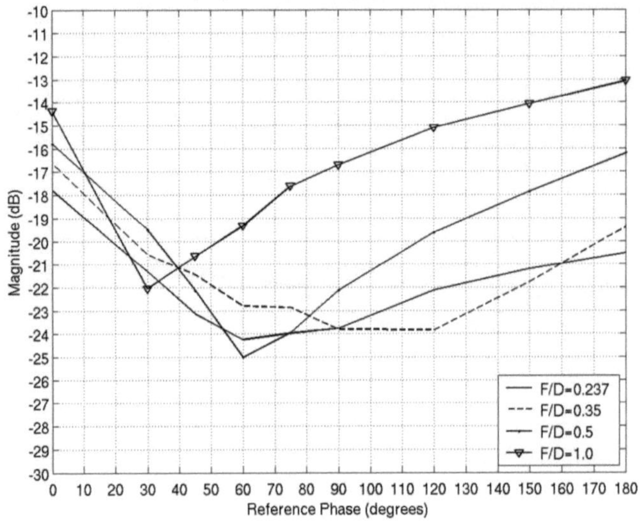

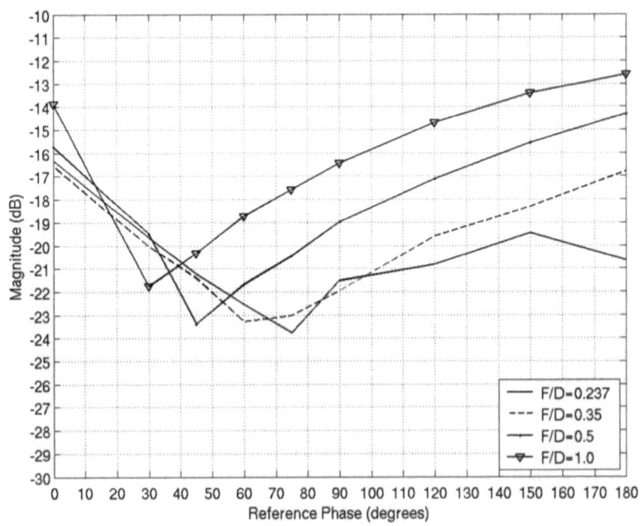

Fig. 2.3 Relative 1st sidelobe level with varying reference phase: E_φ and E_ϑ cut

Table 2.2 and Fig. 2.3 indicate that at least a 6-dB drop in the level of the 1st sidelobe can be expected by changing the reference phase from 0° to a more optimal value. At best this drop could be as much as nearly 9.3 dB, but on average it was around 7.5 dB. The simulations showed that there was no single value of reference phase which would result in the lowest sidelobe for all lens designs. An optimized value must be determined for each value of *F/D*.

The case where the optimum reference phase was 120° *(F/D* – 0.35) was a case where the sidelobe had shifted in relation to the sidelobes formed by the other reference phase cases. This seemed to be a common effect caused by the increase in reference phase. At first, the sidelobes would decrease and would then diminish completely and reform another sidelobe further away from boresight.

Table 2.2 Normalized minimum sidelobe (SLB) level, decrease and reference phase

F/D	H-Plane Min. SLB (dB)	H-Plane SLB decrease (dB)	H-Plane Ref. phase (deg.)	E-Plane Min. SLB (dB)	E-Plane SLB decrease (dB)	E-Plane Ref. phase (deg.)
0.237	–24.22	6.44	60	–23.79	7.47	75
0.35	–23.85	7.23	120	–23.31	6.78	60
0.5	–25.08	9.29	60	–23.46	7.72	45
1.0	–22.08	7.69	30	–21.78	7.89	30

3) Half Power Beamwidth: The 3-dB beamwidth in all *F/D* cases was found to decrease with increasing reference phase. A small part of this was expected due to the increasing diameter of the lens as the reference phase gets larger. However, it was found that the change in diameter could only account for about 0.2° of the overall change, which was typically about 0.5°. Therefore, by increasing the reference phase, a small narrowing in the 3-dB beamwidth can be expected. This effect is visible in Fig. 2.2.

4) Maximum Cross-Polarization: For all cases, changing the reference phase had only a small effect on the cross-polarization levels in the H- and E-planes (Figs. 2.4 and 2.5).

The maximum cross-polarization is fairly constant over all reference phases in the E_ϑ plane. The *F/D*=0.23 case shows a slight increase in maximum cross-polarization with increasing reference phase, but it is only about 5 dB. The E_φ plane curves show that the maximum cross-polarization is certainly affected by the reference phase. Both the *F/D*=0.23 and *F/D*=1.0 show significant increases in maximum cross-polarization at reference phase of 30° and 45°, respectively. It is interesting to note that the *F/D*=0.23 case shows the lowest maximum cross-polarization with reference phase in the E_φ plane, but the highest in the E_ϑ plane.

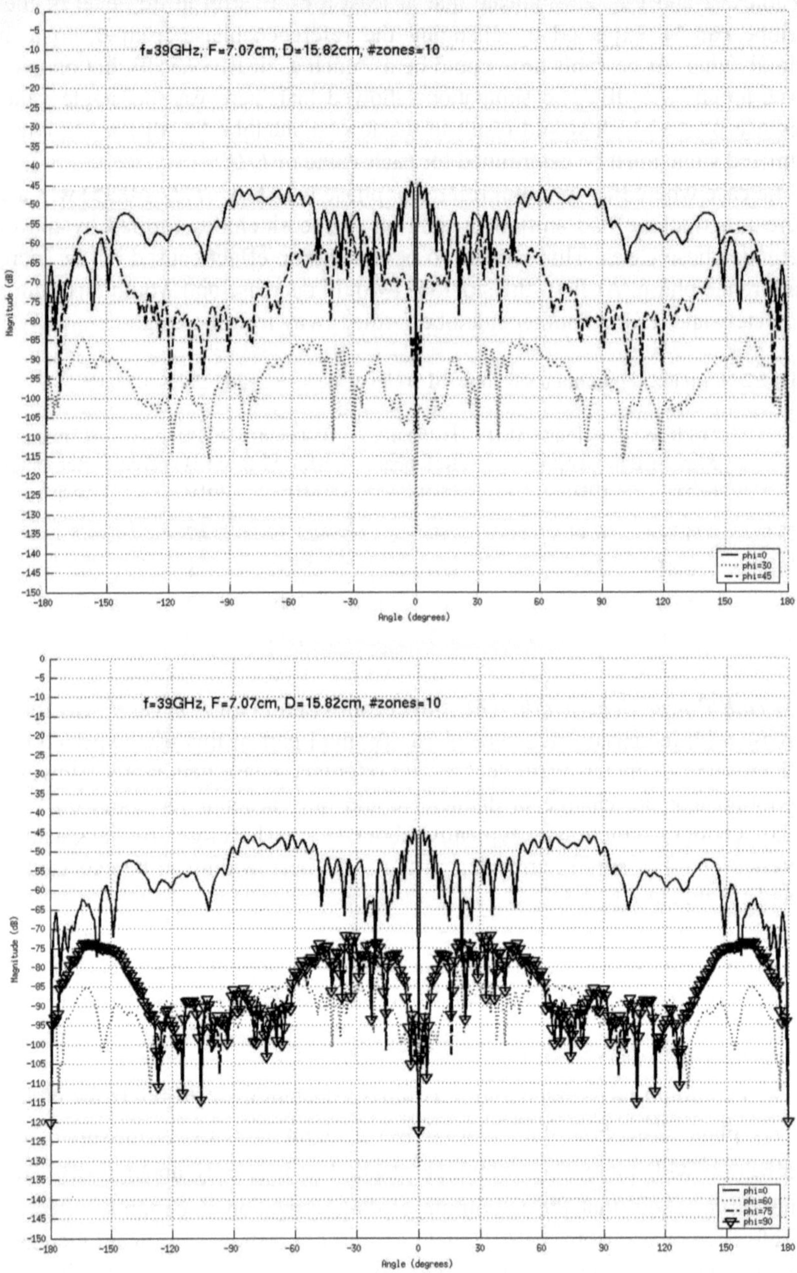

Fig. 2.4 Cross- polarization levels vs. reference phase $\Phi_0 = 0°$, 30°, 45°, 60°, 75°, and 90° in E_φ

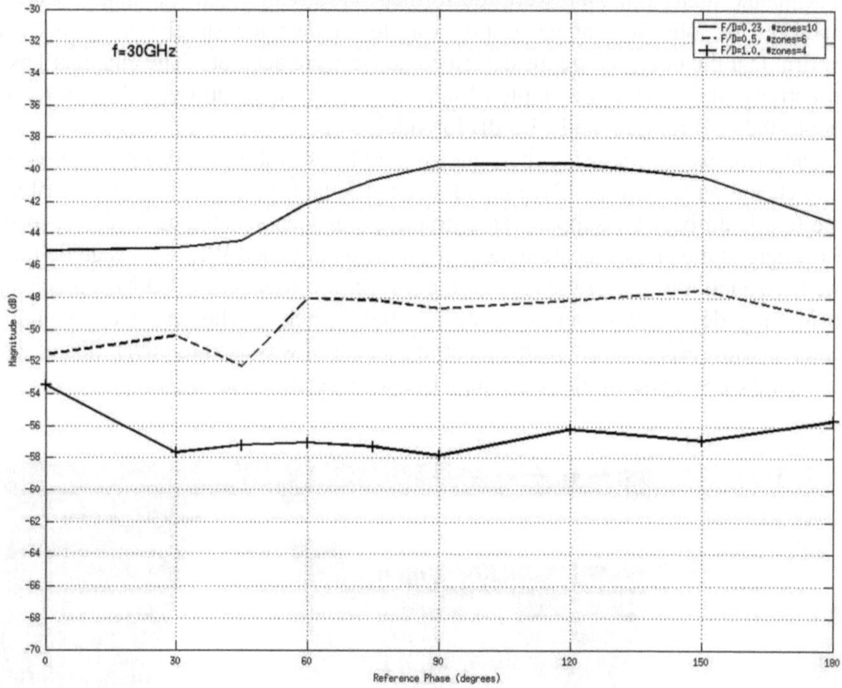

Fig. 2.5 Maximum cross-polarization vs. reference phase E_{ϑ} and *F*/*D*

In order to extract a physical meaning from these simulated results, it is useful to consider the FZPL antenna as a diffractive grating element with a non-periodic structure [2], which is valid for FZP with small number of zones ($N < 100$). Such an element produces output beams having several diffractive orders, the interference of which yields the beam pattern in the far field. Typically, only the −1 order is considered. When the reference phase is altered, a new zone is created, which increases the spatial frequency of the zones in the structure, thus resulting in a somewhat larger lens. A change to the spatial frequency causes a change in the intensity of the various diffractive orders, which redefines the output beam power from the FZPL surface into space. The resulting beam patterns are formed based on this new interference pattern.

Changing the reference phase is thus a means of changing the diffraction and interference patterns of the antenna. Due to the complicated nature of these effects, it is difficult to predict how the radiation patterns of the antenna will be affected without electromagnetic simulations. However, it is intuitive that the antenna's exact geometry is important in forming the diffraction patterns, since constructive and destructive addition of waves will be different in each case thus affecting the sidelobes and locations. This also explains why the optimum reference phase is different for each *F*/*D* antenna configuration.

Antenna measurements were performed in a calibrated far-field anechoic chamber. The objective of the testing was to validate the simulation results and confirm that the observed trends described above were accurate. The chamber has a reflectivity of 30 dB and a ripple of 0.5 dB, which yields an overall accuracy of about ±0.25 dB for gain and ±1.5 dB for sidelobes.

Each lens was made using the same zone radii as in the simulations. The lenses were etched on a 5 mil (0.127 mm) FR4 material and were attached with masking tape to a foam base for support. This structure was then situated on a custom-made test fixture at the proper location away from the open-ended WR28 waveguide feed. The horizontal location of the feed on the test fixture was controlled by a precision positioning device, with an accuracy of ±0.003 inches. The lens was manually centered with respect to the feed. Absorbing material was used to cover all surfaces that could cause undesirable scattering. The setup is shown in Fig. 2.6.

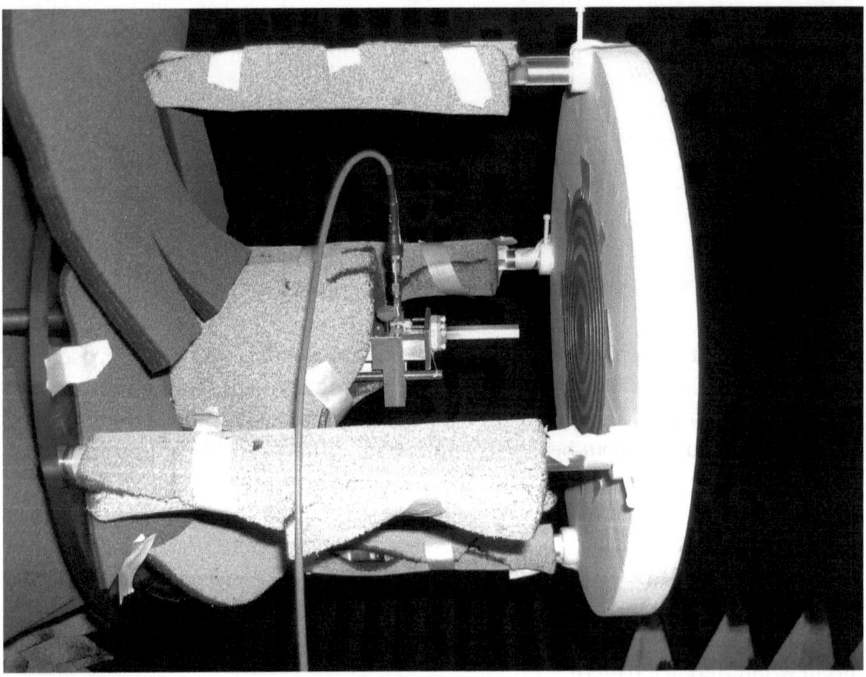

Fig. 2.6 Test setup in the anechoic chamber

A typical measured radiation pattern shows the important trend that the level of the radiation pattern drops with increasing reference phase in the sidelobe vicinity. It is interesting to note that the frequency where the measured peak gain occurred was different from the frequency of the peak gain in the simulations. The simulated frequency of the peak gain remained constant at 30 GHz in all cases with varying reference phase at a given *F/D*, whereas the measured frequencies were 31 GHz for the lower reference phase antennas (0°, 30°, 60°) and then closer to 30 GHz for the 90° and 120° reference phase antennas. The fact that the frequency of the peak gain for the lower reference phase antennas was at 31 GHz could be explained by a number of possibilities including: (1) the frequency of the peak gain for the feed was closer to 31 GHz than 30 GHz, (2) the simulations did not include the FR4 material or the foam support, and (3) the higher gain of the feed at 31 GHz had a narrower beam and thus a smaller edge illumination.

Thus, the measured results revealed that changing the reference phase from the standard 0° to some other value between 0° and 180° caused a drop in the level of the 1st sidelobe between 3 and 7.5 dB depending on the reference phase. These numbers were better in the simulations that showed a drop in the 1st sidelobe between 6.5 and 9.3 dB. This finding is significant since the sidelobe improvement comes with very little impact on either the peak gain or the cross-polarization levels and a small decrease in the 3-dB beamwidth. This trend was also found to exist with other *F/D* values, but the optimum reference phase changes with each antenna configuration. This approach of optimizing the reference phase to improve the sidelobe performance should also be applicable to various types of phase-correcting Fresnel lenses whose radiation efficiencies will be significantly better than the Fresnel zone plate lens.

It could be noted that the concept of beampattern control by means of reference phase is valid not only for normal incident radiation but also for oblique illumination too. For example, on Fig. 2.7 FDTD simulation of the main beampattern parameters of CFZP antenna (source: open circular waveguide at 10°, $F_0 = 150$ mm, $D = 199$ mm, $D/F = 1.33$, $\lambda = 10$ mm, angle of oblique illumination is 10°) are shown.

These results show the classical CFZP is not optimal and beampattern may be optimized vs. reference phase for oblique illumination too.

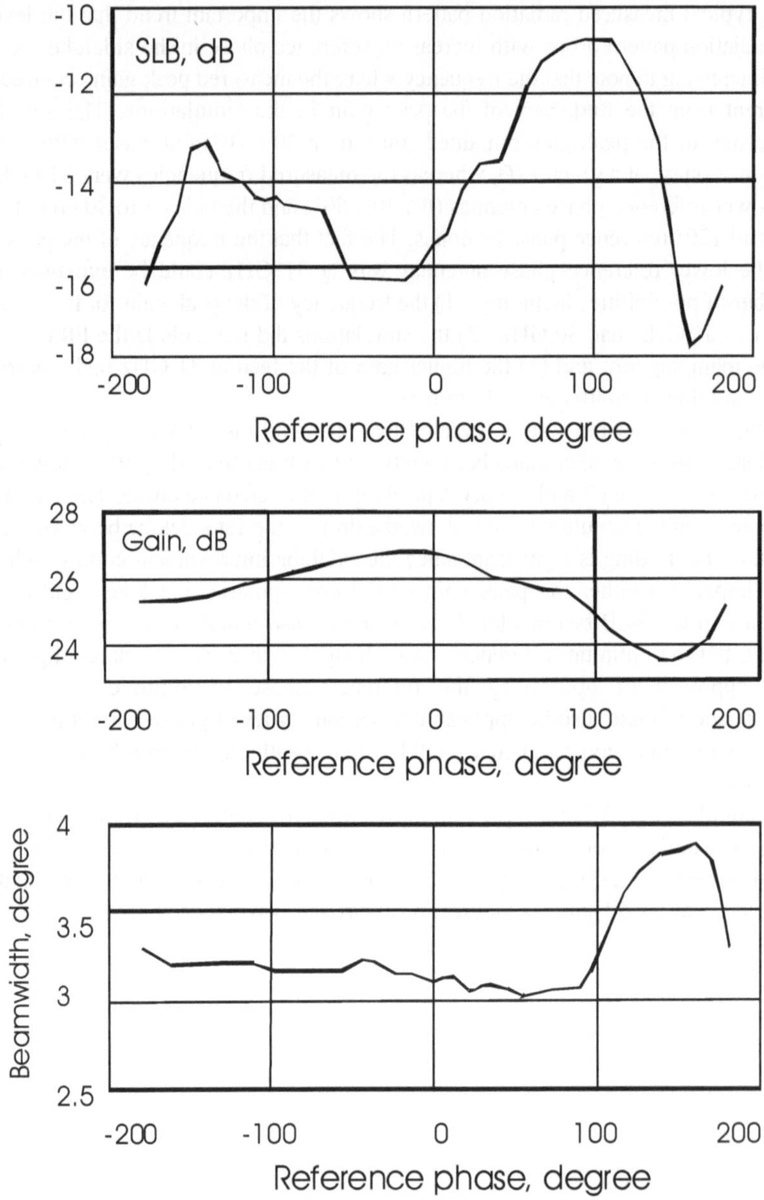

Fig. 2.7 CFZP antenna beampattern parameters vs. reference phase for oblique illumination

2.2 Square FZP

Although symmetry defines the Fresnel zones as circular, for some applications it could be useful to have rectangular or square FZPLs. For example, square FZPLs are better suited to the intrinsic geometry of an array of rectangular pixels. Square Fresnel antenna should be easier to build than equivalent resolution monolithic apertures or classical aperture synthesis arrays, and allows larger fields and high dynamic range for imaging.

Among high angular resolution imaging techniques, aperture synthesis provides the highest resolutions, but requires stringent path lengths control and quasioptical surfaces quality. The previously required high-precision quasioptics are replaced by mere holes in a thin opaque foil covering the dimensions of the array. Focusing is achieved with no other optical element than vacuum and the edges of the holes, their shape and positioning alone being responsible for focalization and beam combining. The consequences of this high number of apertures are a large number of pixels and a very high dynamic range.

Some advantages of square FZP antenna are:

- The angular resolution of a square Fresnel antenna is the same as that of a filled aperture having the size of the whole antenna.
- The use of air (vacuum) for the individual sub-apertures eliminates the phase defects and spectral limitations that would result from a transparent or reflective material. Thus, the spectral span of Fresnel antenna is limited only by the characteristics of the opaque material.
- The resulting wavefront quality is relatively insensitive to array warping and only limited by the precision to which the apertures are carved. This constraint is also quite loose compared to quasioptical surfacing.
- The dense rectangular pattern of the array (all aperture edges following two orthogonal directions) casts light at focus into a central peak flanked by two orthogonal spikes, rather than diffraction rings (case of circular FZP) or broad sidelobes, thus most of the field remains at very low stray light level.
- The large number of apertures allows a high field/resolution ratio, hence much broader fields than with other antenna.

The first attempt to treat the problem to study the field distribution near the focus of a square microwave lens, both theoretically and experimentally, was made in 1956 by P. A. Matthews, A. L. Cullen [3]. Theoretically, they have shown that scalar wave function for the square aperture is

$$\phi = \frac{j\varepsilon^{-jkz}}{\lambda\zeta_0^2}\int_{-\xi_0}^{+\xi_0}\int_{-\eta_0}^{+\eta_0} A(\eta_1\xi)\exp\left[jk\left(\frac{x\eta+y\xi}{\zeta_0}+\frac{\eta^2+\xi^2}{2\zeta_0^2}-z\right)\right]d\eta d\xi$$

The field patterns were measured by means of the spinning-dipole technique. The experimental points measured in the focal was a three-step “egg-box” type with a square aperture 120 × 120 cm. It is two-point corrected for scanning, and was designed to produce a parallel beam. Its focal length was 192 cm.

The transmitting horn was situated at a point twice the focal length from one side of the lens, and the field was measured in a region approximately the same distance from the other side of the lens. The position of the transmitting horn was adjusted to give a symmetrical pattern in a plane transverse to the direction of propagation near the focus and to give as good an approximation to a spherical wavefront as possible.

Particularly satisfactory agreement has been found with the theoretically predicted transverse field pattern in the focal plane, and with the theoretically predicted increase in axial wavelength in the vicinity of the focus. The latter effect is related to the so-called anomalous phase change of 180° on passing through the focus [2]. A phase change of 180° between the main lobe and the first sidelobe in the focal plane, which the theory predicts, has also been verified experimentally.

2.2.1 Gain of a Rectangular Aperture

Gain of a rectangular aperture in the far field was considered in [4]. It was shown the gain of rectangular aperture in the Fraunhofer region is:

$$G = G_0 \frac{C^2(u_1)+S^2(u_1)}{u_1^2}\frac{C^2(p_1)+S^2(p_1)}{p_1^2} \text{ or } G = G_0 M(u_1)M(p_1), \tag{2.1}$$

where $M(u)=[C^2(u)+S^2(u)]/u^2$ and $u_1 = L_1/\sqrt{2R\lambda}$, $p_1 = L_2/\sqrt{2R\lambda}$.

For a square aperture $L_1 = L_2 = L$, and accordingly we let $u_1 = p_1 = u$ and Eq. (2.1) becomes

$$G_{\text{sq}} = G_0 M^2(u).$$

The corresponding equation for the circular aperture was

$$G_{\text{cir}} = G_0 \left(\sin(x)/x\right)^2.$$

On other words, for rectangular aperture with *uniform distribution of amplitude and phase* of the radiation field, the distribution of field at the far zone can be represented as following [2]:

$$U(\theta,\varphi) = \frac{\sin[(\pi a/\lambda)\sin\theta\cos\varphi]}{(\pi a/\lambda)\sin\theta\cos\varphi}\frac{\sin[(\pi b/\lambda)\sin\theta\sin\varphi]}{(\pi b/\lambda)\sin\theta\sin\varphi}. \tag{2.2}$$

It is followed from Eq. (2.2) that directional patterns in principal planes (i.e. in planes xz or yz) are identical. For diagram in plane xz value $(\pi a\sin\theta)/\lambda$ is used as an argument, and in plane yz value $(\pi b\sin\theta)/\lambda$ correspondingly.

Radiation field in the case of circular aperture is:

$$U(u) = 2\pi a^2 J_1(u)/u. \tag{2.3}$$

Two differences are immediately apparent:

1) Near the outer limit of the Fresnel zone, the correction for the square aperture leads to a smaller net gain than the correction for the circular aperture.
2) The gain correction factor for the circular aperture has zeros at $x = \pi n$ while the rectangular aperture has minima, but not zeros at $x =$ 3.65, 7.62.

Beampattern (BP) by power at a distant zone (i.e. field intensity) is proportionate to a squared field's amplitude $|U(u)|^2$. For a principal plane with rectangular aperture, BP fits the expression $[\sin u/u]^2$ while for circular aperture – $[J_1(u)/u]^2$.

For a circular aperture, a beamwidth at a half power level (~ 3 dB) is equal to arcsin (1,029λ/D) and corresponds to an arguments value *u=1,616.* For a rectangular aperture, a beamwidth is equal to arcsin $(0{,}887\lambda/l)$ and corresponds to an arguments value *u=1,393.* At the principal plane, a first sidelobe is 13.2 dB less than the main lobe for rectangular and 17.5 dB less than a main lobe for circular apertures.

Values of the first four extremes of the filed intensity distributions for circular and rectangular apertures with uniform distribution of amplitudes and phases by aperture are shown at the following Table 2.3.

It also was shown [3] that L = 1.67R by *equating the gain* of the circle with the gain of the square aperture.

Table 2.3 Four extremes of the filed intensity distributions for circular and rectangular apertures

Circular aperture		
Points of maximum	Points of maximum	Relative intensity
$\sin\varphi_1 - 0.61\lambda/r$	$\sin\varphi = 0$	1.00
$\sin\varphi_2 = 1.12\lambda/r$	$\sin\varphi = 0.81\lambda/r$	0.0175
$\sin\varphi_3 = 1.62\lambda/r$	$\sin\varphi = 1.33\lambda/r$	0.0042
$\sin\varphi_4 = 2.12\lambda/r$	$\sin\varphi = 1.85\lambda/r$	0.0016
Rectangular aperture		
$\sin\phi = n\lambda/b;\ n = 1,\ 2,\ ...$	$\eta = \pi b\sin\varphi/\lambda$	
λ/b	0	1.00
$2\lambda/b$	$\eta = 1{,}43\pi$	0.047
$3\lambda/b$	$\eta = 2.46\pi$	0.008
$4\lambda/b$	$\eta = 3.47\pi$	0.005

2.2.2 Diffraction Efficiency

Let us define the "diffraction efficiency" of a Fresnel antenna as the proportion of incident radiation going into the first-order focus. It is well known that the theoretical diffraction efficiency η_c for a circular Fresnel zone plate when the number of zones tends to infinity is: $\eta_c \approx 1/\pi^2 m^2$ for m odd; and $\eta_c \approx 0$ for m even and $m \neq 0$, where m is the diffraction order.

At the prime focus, which corresponds to the diffraction order $m = 1$, about 10.1% of the radiation is gathered.

In the case of the square FZP, the diffraction efficiency can be derived using Cartesian coordinates. The wave contribution at focus can be expressed:

$$E \approx A_0 \int_{rect} \exp(i\phi(x,y))\mathrm{d}x\mathrm{d}y = A_0 \int_{\rho_x}\int_{\rho_y} \exp\left(i2\pi\frac{mx}{p_x}\right)\exp(i2\pi\frac{my}{p_y})\mathrm{d}x\mathrm{d}y,$$

where approximation of $\varphi(x,y) \approx 2\pi\left(\frac{m_x}{\rho_x}+\frac{m_y}{\rho_y}\right)$ was used because the phase within a given aperture is a quadratic function of position and $N >> 1$.

For m odd and $m \neq 0$, the diffraction efficiency in amplitude for a given element of a square Fresnel antenna can be expressed as:

$$\frac{E_{\max}}{E_0} = \frac{4}{\pi^2 m^2},$$

and considering a 50% overall "void to total area" transmission ratio, the diffraction efficiency in amplitude is $\frac{2}{\pi^2 m^2}$. Finally, the diffraction efficiency in intensity for a square Fresnel element is:

$$E_s = \frac{4}{\pi^2 m^2}.$$

When apodization is not applied, at order $m = 1$ this corresponds to a 4.1% efficiency.

For example, two separate orientations of the square FZPs are shown in Fig. 2.8: one where the FZP is oriented vertically (Case 1) and one with the FZP oriented diagonally (Case 2). This was done to determine the effect of polarization on the pattern performance. Figure 2.9 shows the patterns for the square FZP with traditional zoning rules.

The gain for the square FZP with vertical orientation is about 16.7 dB, but for diagonal orientation is 18.2 dB. The first sidelobes for square with vertical orientation is about –3.7 dB and for diagonal orientation is –11.2 dB.

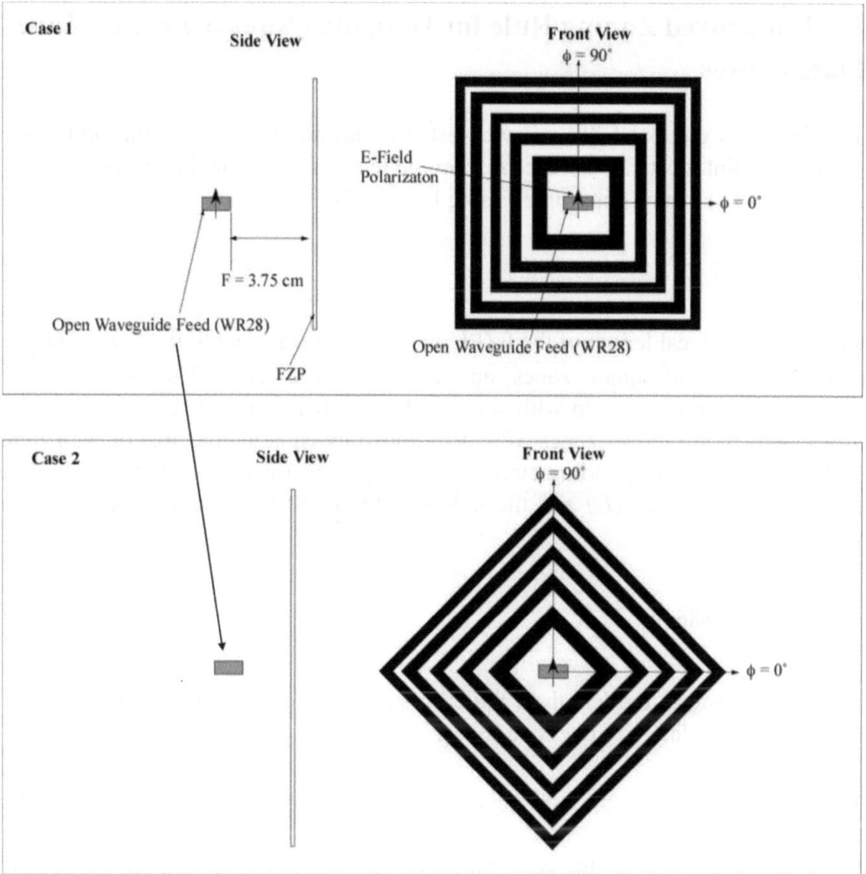

Fig. 2.8 Geometry of the square FZPs

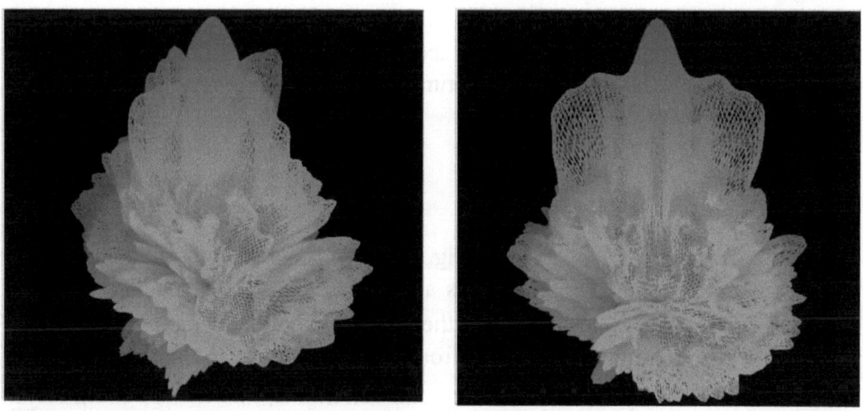

Fig. 2.9 3D beampatterns of the square FZPs: vertical orientation (*left*) and diagonal orientation (*right*)

2.2.3 Improved Zoning Rule for Designing Square Fresnel Zone Plate Lenses

The design of circular FZPLs is well established; and to focus an incident plane wave to a point located a distance F from the lens, as a rule the dimensions of each zone radius (r_i) can be determined using Eq. (1.17):

$$r_i = \sqrt{i\lambda F + (i\lambda/2)^2}\,, \tag{2.4}$$

where F is the focal length of the FZPL and λ is the wavelength in the medium.

For the case of square zones, approximations are required since the square boundaries do not coincide with the circular Fresnel zones. In earlier work, the dimensions of the square zones were determined by equating the area of each zone with that of the corresponding circular zone [5]. Thus the relation between half of the side of the square (L_i) and the radius of the corresponding circular zone (r_i) was [5]:

$$L_i = \frac{r_i\sqrt{\pi}}{2} = 0.8862 r_i\cdot \tag{2.5}$$

In other designs, the length of the each square zone is equated to the diameter of the corresponding circular zone [6, 7]:

$$L_i = r_i\,. \tag{2.6}$$

To derive the new zoning rule [8], based on minimization of a phase error, the geometry in Fig. 2.10 is considered, where the circle of radius R represents the boundary of a particular Fresnel zone centered at point O in the X-Y plane and having a focal point located on the Z-axis at a distance F from the lens. For any point B located on the circle, the following relation is precisely fulfilled within the geometric optics approximation for an axisymmetrical zonal plate (thus we shall consider that these boundaries are optimum for the appropriate axisymmetrical diffractional element):

$$|OB| + |BF| = |OF| + n\lambda/2 \tag{2.7}$$

The solid squares (Q_1 and Q_2) in Fig. 2.10 represent two possible choices for the square FZPL zone. If A represents a point on the square boundary and B a point on the circular zone boundary, then one can define a maximum error Δ between the difference of a set of rays from the lens aperture to the focal point as:

$$\Delta = |OA| + |AF| - \left(|OF| + n\lambda/2\right) \neq 0\,. \tag{2.8}$$

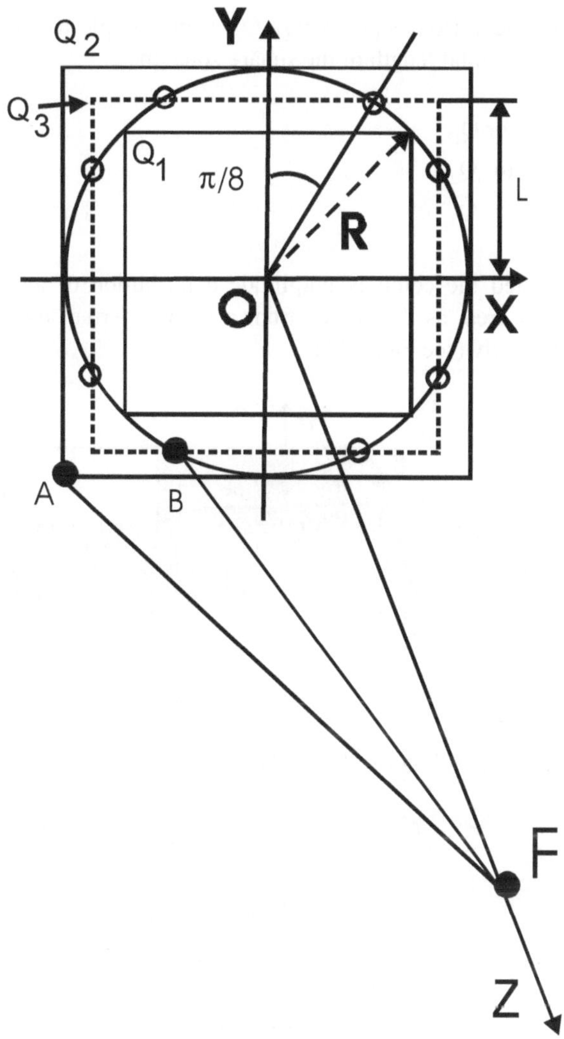

Fig. 2.10 Geometry of a single zone for deriving the improved zoning rule for a square FZPL

The task of optimizing the square FZPL zones is one of minimizing this error Δ. Using a simple geometrical interpretation, this error can be minimized by maximizing the number of common points between the square and the circular zones. The two solid squares in Fig. 2.10 each share only four points with the circular boundary. Choosing a square whose dimensions lie between the two solid squares (the dashed square labeled Q_3 in Fig. 2.10) will increase the number of common points from four to eight and should thus reduce the error. The minimum error will occur when these eight points are equally spaced around the circular zone. The dimensions of the square zone can be related to the radius R of the circular zone for the various squares (Q_1–Q_3) using simple trigonometry. If L is defined as half of the total length of the square zone, then

$$L = \begin{cases} R/\sqrt{2} & \text{, for } Q_1 \\ R & \text{, for } Q_2 \\ R\cos(\pi/8) & \text{, for } Q_3 \end{cases} \tag{2.9}$$

It could be noted the complex amplitude contribution of an aperture is not proportional to its area, as the waves from a given aperture element do not all interfere constructively (see Fig. 2.11).

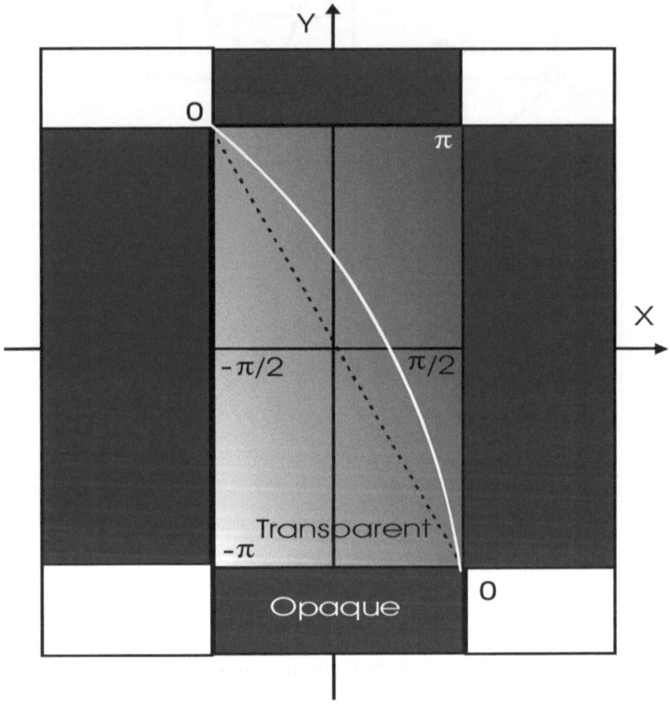

Fig. 2.11 Evolution of the phase within an individual aperture, as seen from common focus

The areas of these three square zones are compared to the area of the circular zone in Table 2.4. It is important to notice that this proposed technique for the definition of the width of the square zone does not depend on a type of the diffractional element (e.g. transformation of a flat or spherical wavefront set, or the focusing to a point or ring) and has a universal character.

Table 2.4 Parameters for various zoning rules

Type of FZPL	Area of zone	Ratio S/S_0
Circular	$S_0 = \pi R^2$	-
Equal Area Square[1]	$S = \pi R^2$	1.0
Interior Square (Q_1)	$S = 2R^2$	$2/\pi = 0.637$
Exterior Square (Q_2)	$S = 4R^2$	$4/\pi = 1.273$
Optimal Square (Q_3)	$S = 4R^2\cos^2(\pi/8)$	$4\cos^2(\pi/8)/\pi = 1.087$

To help verify the improved performance of this new zoning rule, several simulations were carried out using a finite-difference time domain (FDTD) method [8]. The case presented here uses the square FZPL as an antenna, where the FZPL is illuminated by an open rectangular metal waveguide (WR28) as shown in Fig. 2.8 case 1, designed to operate at 30 GHz. The aperture of the waveguide is placed at the focal point of the FZPL, which is located at a distance F from the center of the FZPL. Two lenses with 10 zones were analyzed: one with the $L = R$ zoning rule (Q_2 in Fig. 2.10) and one with the new zoning rule $L = R\cos(\pi/8)$ (Q_3 in Fig. 2.10) where the values of the circular zone radii were determined using Eq. (2.4) and F was chosen to be 37.5 mm. The simulated far-field patterns in the two principal planes ($\varphi = 0°$ and 90° as indicated in Fig. 2.8 case 1) are shown in Fig. 2.12 for these two lenses at 30 GHz. For this case, the new zoning rule improves the peak gain by 1.2 dB over the traditional zoning rule, even for a binary amplitude-type of FZP. Simulations where the square FZP is used as a lens to collimate an incident plane wave show that the new zoning rule also enhances the focusing properties by moving the focal point significantly closer to the lens aperture, compared to the traditional zoning rules.

The following could be mentioned. The derivation in Section 1.2.1 "Gain of a Rectangular Aperture" is based on circular and square apertures that have uniform amplitude and phase distributions, and $L = 1.67R$ is obtained by *equating the gain of the circle with the gain of the square aperture*.

In our derivation, we do not assume uniform amplitude and phase distribution over the apertures. Also, our equation ($L = 0.886R$) is based on maximizing the performance of the square FZP; i.e. the square FZP will not have the same gain as the circular FZP in our case.

This new zoning rule may be used for square FZPs in microwave, millimeter, and terahertz waves, optical and IR techniques for optimal design of diffractive elements with square zone topology, including antennas, lenses, and scanning axicon, etc. This is a familiar fact that when a cylindrical zone plate is illuminated with a plane wavefront, a focal region is a line situated transversely to the optical axis, in parallel to the plane of the zone plate. Therefore, in contrast to the Fresnel zone plate, a "square" zone plate composed of two cylindrical zone plates produces a focusing region of an asymmetrical shape [9]. The focal area has the shape of a "cross" with a maximum in the intersection area. Along the diagonal of the cross, the field intensity distribution is similar to that produced by the Fresnel zone plate. The detailed investigations of square FZP as a lens were described in [2].

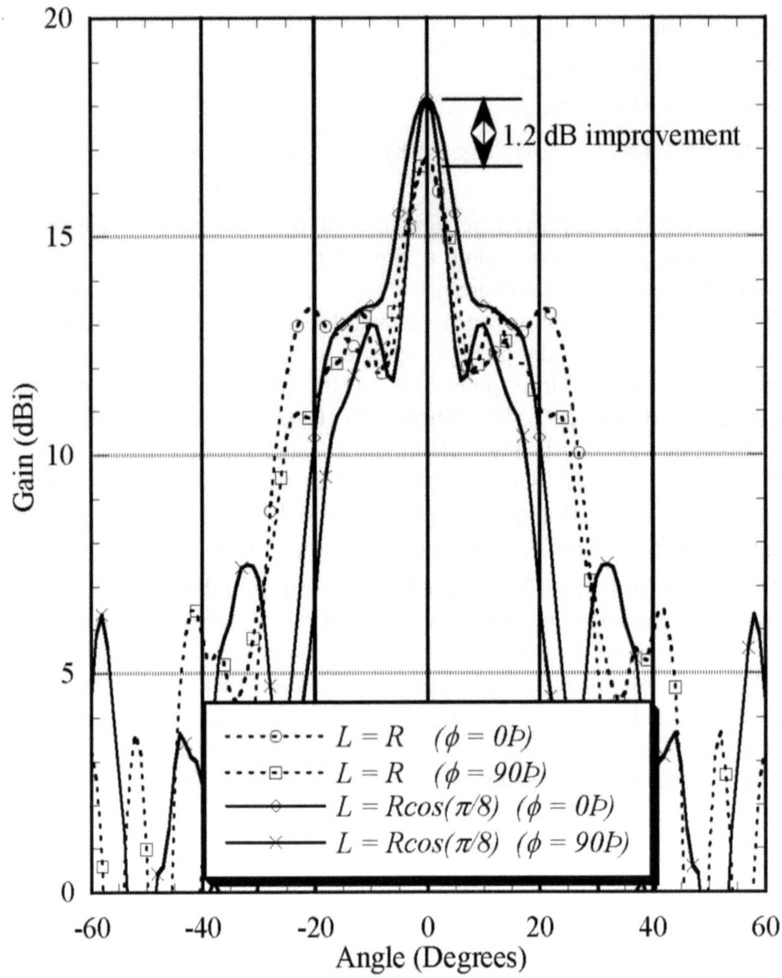

Fig. 2.12 Simulated gain of the FZLPs at 30 GHz

The peak gain vs. frequency for the circular and square FZPs are plotted in Fig. 2.13. For the case of the vertical orientation, the FZP with the optimized zoning has a higher gain than the FZP with traditional zoning (18.2 dB vs. 16.8 dB) at 30 GHz. However, for the diagonal orientation, both the traditional and the optimized FZPs achieve the same gain (17.8 dB) at 30 GHz. Looking at the trend over frequency, the gain of the optimized FZP increases with frequency while the traditional zone FZP has a decreasing gain with frequency.

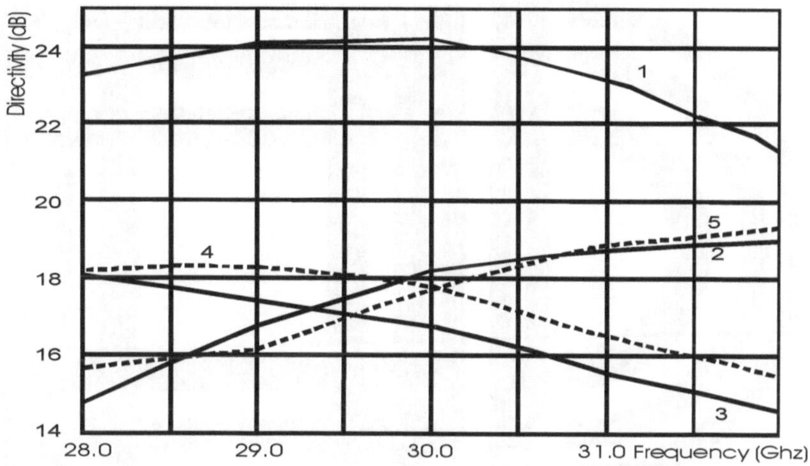

Fig. 2.13 The peak gain vs. frequency: 1 – circular FZP, 2 – square FZP with optimal zone vertical orientation, 3 – square FZP with classical zone vertical orientation, 4 – square FZP with classical zone diagonal orientation, 5 – square FZP with optimal zone diagonal orientation

2.2.4 Fresnel Diffractive-Interferometric Arrays

Some modification of a square FZP lens is possible. Let us consider a design of Fresnel diffractive-interferometric arrays from a circular FZP (Fig. 2.14). The term "Fresnel zone" defines an area delimited in the aperture plane by two concentric circles in the intersection of the aperture plane with spherical waveplanes centered on the focus and whose radii differ by one wavelength. The central Fresnel zone is the disc delimited by the smallest intersection. The number of zones covered by a Fresnel array (as for a filled aperture) corresponds to the number of zones crossed from center to edge along a one-dimensional line.

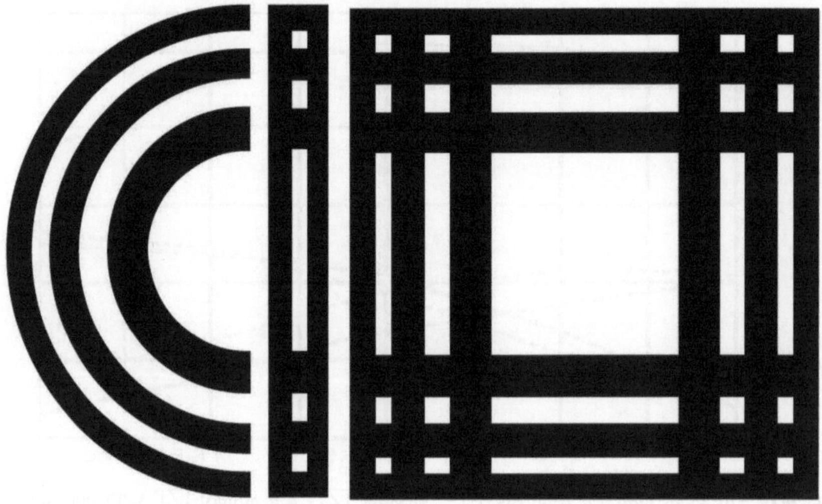

Fig. 2.14 Circular FZP (*left*) and examples of one- and two-dimensional (*right*) orthogonal Fresnel array

As it was mentioned, such orthogonal FZP can be seen either as an aperture synthesis array or as a particular case of classical diffractive zone plate. Beams from the individual apertures are recombined by diffraction and interference. The rectangles apertures (in this case) are positioned so that at the first order of diffraction (2π phase shift from one aperture sub-set to the next), an incoming plane wave is turned into a spherical outgoing wavefront. Their orthogonal layout makes the nontransmissive zones connected over the whole array.

One possible transmission law $T(x, y)$ of the array is built as follows. Let us define functions g as (taking in mind the concept of reference phase):

$$g(x,y)=\begin{cases}1 & \text{if } \sqrt{(x,y)^2+f^2}-\sqrt{(x_0,y_0)^2+f^2}\in\left|\left(k+\frac{f}{m\lambda}+\frac{1}{2}\right)m\lambda;\left(k+\frac{f}{m\lambda}+1\right)m\lambda\right|,\\ 0 & \text{otherwise}\end{cases}$$

where (x,y) are the distances from the optical axis, m is the diffraction order, k the Fresnel zone index, f the designed focal length, and (x_0,y_0) are the arbitrary values of reference radii (in the case of binary array). The limiting condition for the reference radii is:

$$0\leq(x_0,y_0)\leq\sqrt{\lambda f}\,.$$

Also it could be noted that in common case the value of x_0 is independent from the value of y_0. Binary type of rectangle diffractive-interferometric arrays is shown in the Fig. 2.15 [9].

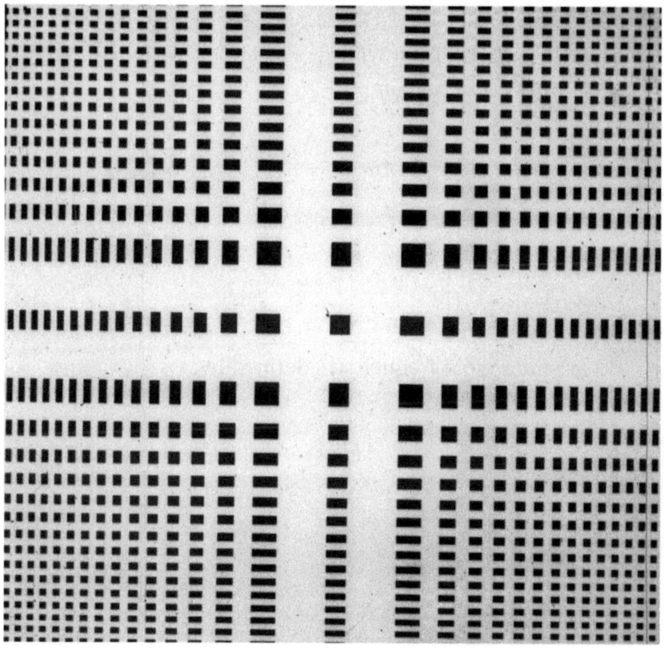

Fig. 2.15 Binary-type of rectangle diffractive-interferometric arrays

Using functions $g(a)$ and $h(a)=1-g(a)$, one can define a two-dimensional transmission based on an orthogonal development of g and h:

$$T_0(x,y)=h(x)h(y)+g(x)g(y).$$

The Point Spread Function (i.e. field intensity distribution in the focal plane) of such arrays is the square modulus of the Fresnel transform of $T(x, y)$.

To simplify a problem and understand a physics of array design, let us consider Yong's double-phase aperture interference (Fig. 2.16). Let us consider a double-square phase aperture of width l. In this case, the diffracted wave field at point P is [10]:

$$U(P)=U_G+[1-\exp(ik\Delta_1)]\,U_{BA}+[1-\exp(ik\Delta_2)]\,U_{BA'}, \tag{2.10}$$

where

$$U_G=\begin{cases} A\exp(ikz), & P\in L \\ A\exp(ik(z+\Delta_1)), & P\in L_A \\ A\exp(ik(z+\Delta_2)), & P\in L_{A'} \end{cases}$$

and

$$U_{BA} = K\exp(-i\pi f_x d)l^2 \frac{\sin(\pi l f_x)}{\pi l f_x}\frac{\sin(\pi l f_y)}{\pi l f_y},$$

$$U_{BA'} = K\exp(i\pi f_x d)l^2 \frac{\sin(\pi l f_x)}{\pi l f_x}\frac{\sin(\pi l f_y)}{\pi l f_y}.$$

$$K = -\frac{1}{2\pi\lambda z}\exp(ikz)\exp\left(ik\left[\frac{x_0^2+y_0^2}{2z}\right]\right).$$

Here (f_x, f_y) are the spatial frequencies defined by:

$$f_x = \cos(\theta/\lambda) \approx x_0/\lambda z,\ \ f_y = \cos(\varphi/\lambda) \approx y_0/\lambda z.$$

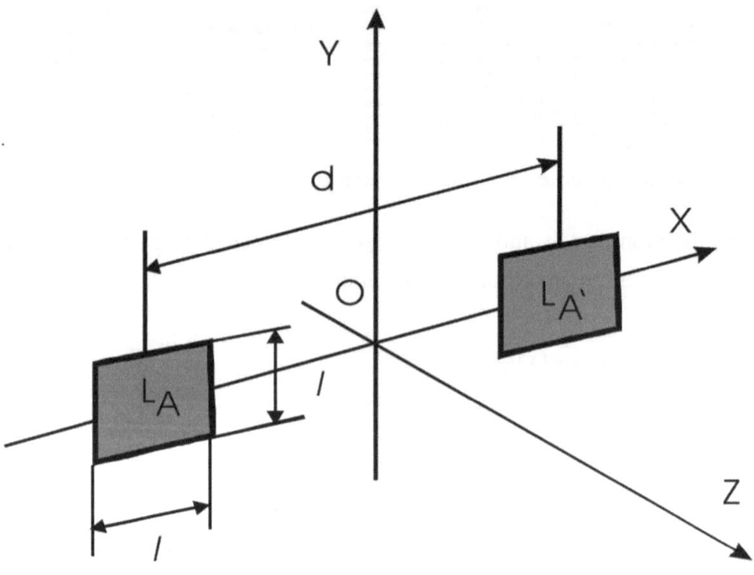

Fig. 2.16 Double-phase rectangle aperture

Simple inspection of Eq. (2.10) shows that one does not have a diffracted field whenever $\exp(ik\Delta_1) = \exp(ik\Delta_2) = 1$. If $\exp(ik\Delta_1) = \exp(ik\Delta_2) = \exp(ik\Delta)$, the boundary wave contribution to the diffracted wave field is:

$$U_B = K(1-\exp(ik\Delta))\, l^2 \frac{\sin(\pi l f_x)}{\pi l f_x}\frac{\sin(\pi l f_y)}{\pi l f_y}\cos(\pi f_x d). \tag{2.11}$$

Aside from the amplitude and phase factors, this expression of the boundary wave is the classical result of the Fourier-optics treatment of Young's double-slit experiment [11]. A simple inspection of Eq. (2.11) shows a maximum of fringing effect in the diffracted wave field when $\exp(ik\Delta) = -1$.

2.2.5 Polygonal FZP

2.2.5.1 Geometrical Optimization of Polygonal Fresnel Zone Plates in Paraxial Approximation

When the polygonal shape is used in an array, as shown in Fig. 2.17, it does not result in overlap issues like the circular geometry. However, since the polygonal FZPL is inherently an approximation to the CFZPL, it will not perform as well. In order to make the polygonal FZPL a more attractive candidate for use in an array, new methods of the CFZPL approximation by polygons, for instance, introduced in [8] for a square Fresnel zone plate antenna, could be developed.

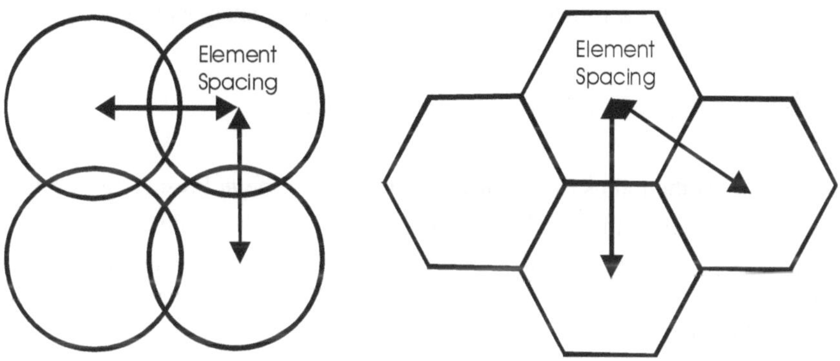

Fig. 2.17 Arrays of circles (*left*) and hexagons (*right*)

Moreover, when the resolution of the fabrication techniques is comparable to the spatial dimensions of the Fresnel zones, the actual fabricated contours circles are jagged versions of the perfect circular pattern. For those cases, the study of polygons with a large number of sides has received attention [12, 13].

As it was mentioned, a very useful approximation neglects the second term within the square root and provides the well-known dependence where the zone radius is proportional to $\sqrt{m}$. This is true in optical waveband and these values serve as a reference when calculating the geometric parameters of the polygon element in [14] (it means a paraxial approximation, which is not valid in microwave [2]).

The circle of reference for the zone m has a radius R_m, given by Eq. (2.10). The parameter for the polygonal shape will be the apothem x, along with the number of sides s. The optimum polygon contour should intersect the circle of reference at a position that overfills the circle of reference in some regions and underfills it in the rest. When analyzing regular convex polygons, we may restrict the region of computation to an angular range determined by an apothem and the line joining the center of the polygon and the closest vertex to the apothem location. The angle subtended by these lines is

$$\alpha_{\text{pol}} = \pi / s,$$

where s is the number of sides of the polygon. The region of the polygon within this angular range always has the shape of a right triangle. There are $2s$ equal right triangles in a regular polygon having s sides. From Fig. 2.17, we may see that the polygon shape intersects with the circular Fresnel zone reference only once on the contour side of this right triangle, at an angular position given by α_{int}. The location of this intersection is the main issue that needs to be properly solved. By reducing the area that overfills and underfills the circular angular sector when the polygon is traced, the authors of [14] decrease the number of Huygens wavelets that belong to the areas A_1 and A_2 in the Fig. 2.18.

The extent of this mismatched area has the following analytical form,

$$A = A_1 + A_2 = R_m^2 \left(\alpha_{\text{int}} - \frac{1}{2}\alpha_{\text{pol}} \right) + x^2 \left(\frac{1}{2}\tan\alpha_{\text{pol}} - \tan\alpha_{\text{int}} \right),$$

where R_m is the radius of the circular boundary of the m Fresnel zone, x is the value of apothem of the regular polygon, and α_{int} is the angle that defines the intersection between the polygon and the circular contours (as seen in the lower portion of Fig. 2.17). This value can be related to the apothem value by the following relation,

$$\alpha_{\text{int}} = \cos^{-1} \frac{x}{R_m}.$$

By minimizing the total area A with respect to the value of the apothem x, we find the optimized apothem value,

$$x_m = \frac{R_m}{\sqrt{1 + 0.25\tan^2(\pi / s)}}.$$

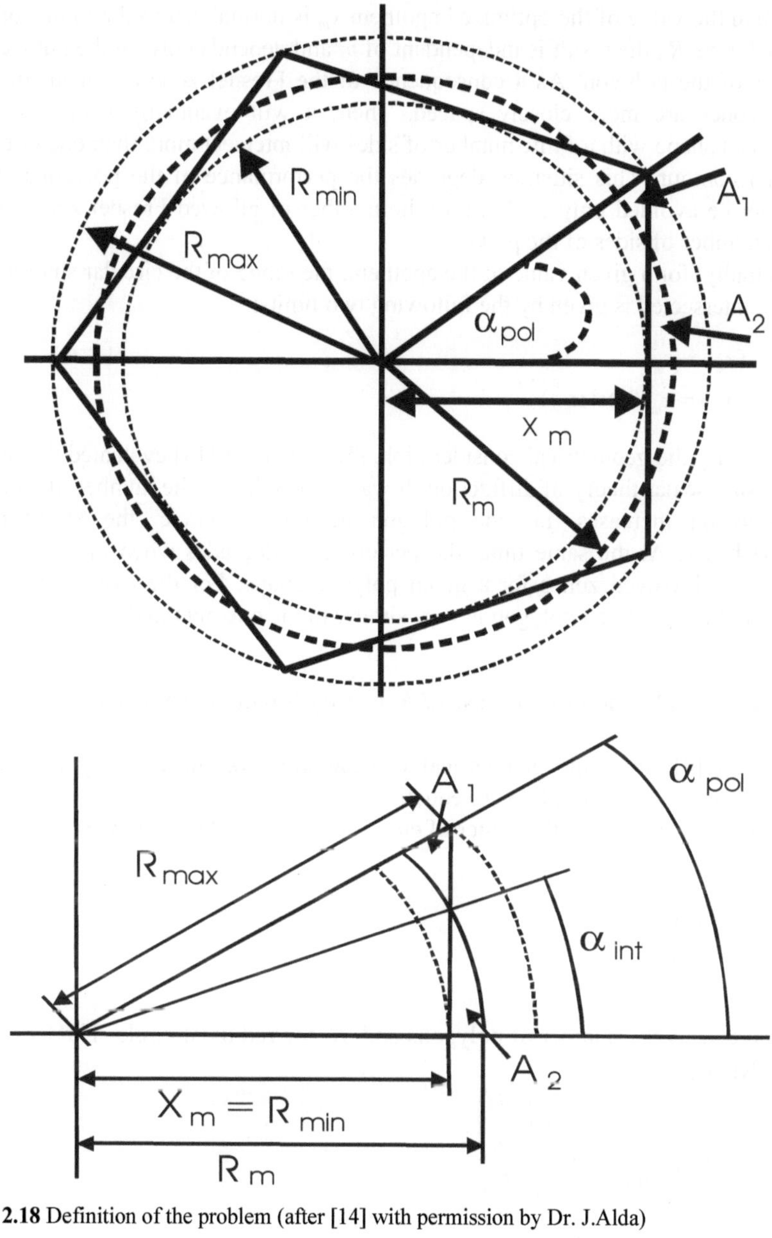

Fig. 2.18 Definition of the problem (after [14] with permission by Dr. J.Alda)

When the value of the optimized apothem x_m is normalized to the radius of the Fresnel zone R_m the result is independent of m and depends only on the number of sides s of the polygon. As a consequence of the Fresnel zone arrangement, the outer zones are more closely spaced. Then, it will eventually happen that a polygonal shape with a finite number of sides will intersect more than one circular Fresnel contour. This situation degrades the performance of the polygonal FZP and can be avoided only by limiting the number of allowed Fresnel zones for a given number of sides of the polygon.

Actually, for a given value of the apothem, the range of the circular shapes that can be intersected is given by the following two limits:

$$\begin{aligned} R_{\min}(x_m) &= x_m \\ R_{\max}(x_m) &= x_m / \cos(\alpha_{\text{pol}}) \end{aligned}.$$

To verify the geometrical consideration, the authors of [14] evaluated the Strehl ratio using scalar theory of diffraction. It was shown that as the number of sides of the polygon increases, i.e. the polygon becomes a circle, the Strehl ratio approaches 1. At the same time, the performance degrades when increasing the number of involved zones for a given polygon shape. For the case of a square polygon, the optimized polygon is very similar to the one obtained recently in [8].

2.2.5.2 Generalization to the Case of Arbitrary Shape of Aperture

Following [4], from the geometrical consideration for the arbitrary polygonal shape we may write the following equations.

The distance between the center of circle and polygonal boundary to the normal is:

$$x = RCos\left(\frac{1}{2}\frac{2\pi}{2n}\right),$$

where n is the number of polygons and R the radius of circle. The length of the polygons is:

$$L = 2R\sin\left(\frac{\pi}{2n}\right) + 2R\frac{\sin\left(\frac{\pi}{2n}\right)}{\sin\left(\frac{\pi}{n}\right)}.$$

For the optimal hexagonal configurations, the distance from the center of the hexagon to the center of an edge L will be related to the radius of the circle R by the expression in equation:

$$L = 0.996R.$$

The following could be noticed. The optimization proposed by J.Alda et al. described in Sect. 2.3.1 "Geometrical optimization of polygonal Fresnel zone plates in paraxial approximation" was based on the geometrical minimization of the area that overfills and underfills the circular angular sector when the polygon is traced. The optimization based on the reduction of the phase mismatch between the contribution to a given pseudo-Fresnel zone of a polygon and an actual circular Fresnel zone [8] described above is more physically clear and correct.

2.2.5.3 Hexagonal FZP vs. Reference Phase

As an example let us briefly consider the beampattern analysis of HFZP vs. reference phase. Parameters of the simulation were: lens shape – hexagonal with optimum dimensions, number of zones: 10, F/D: 0.23 (F = 3.75 cm, D ~ 15.8 cm), frequency: 30 GHz, reference phases: 0°, 30°, 45°, 60°, 75°, 90°, 120°, 150°, 180°.

a) *Co-polarized Radiation Patterns*: Figure 2.19 shows the E_φ and E_ϑ cuts, respectively, of the co-polarized radiation patterns. As the reference phase increases in the E_φ cut, the gain drops, the first null gets deeper, and the first sidelobe peak level decreases. There is a clear improvement in the first sidelobe level by choosing a reference phase other than 0°. The situation is somewhat different in the E_ϑ cut where although the gain decreases similarly, the first sidelobe level increases with increasing reference phase. Also, the location of the first null and first sidelobe peak varies with reference phase. When the reference phase is 30°, the E_ϑ behavior is significantly different from the other cases.

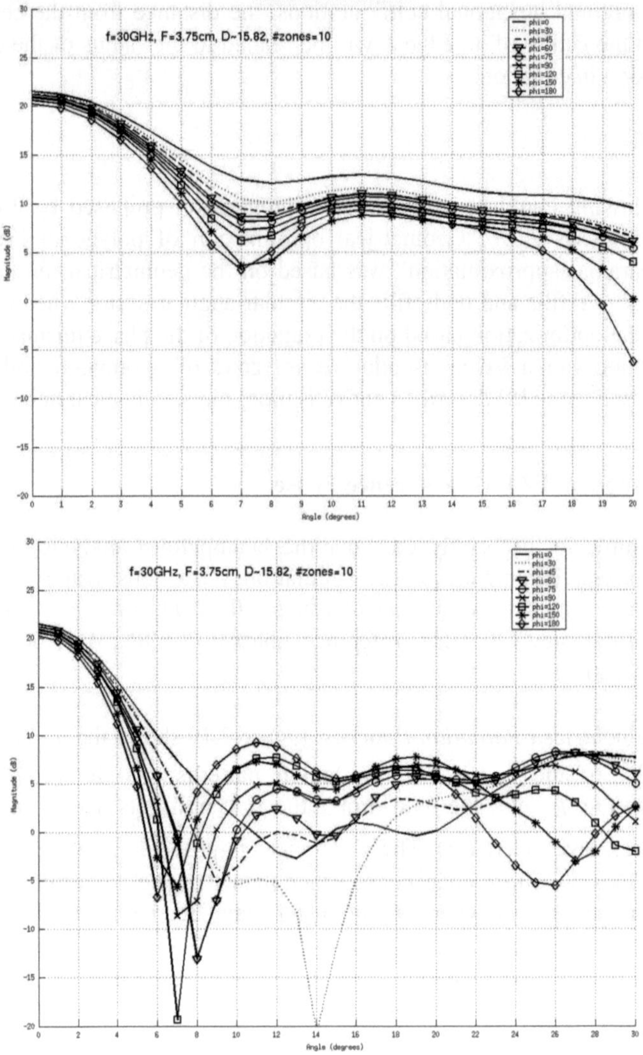

Fig. 2.19 Co-polarized radiation patterns: E_{φ} and E_{ϑ} cuts

Figures 2.20 shows the three-dimensional co-polarized radiation patterns for optimal HFZPL antenna.

Fig. 2.20 Three-dimensional Pattern of the optimized HFZP at 30 GHz

b) *Peak Gain*: Figure 2.21 illustrates the drop in peak gain as seen in the previous two figures. The overall drop is about 1.3 dB. Results are the same for both E_{φ} and E_{ϑ} cuts.

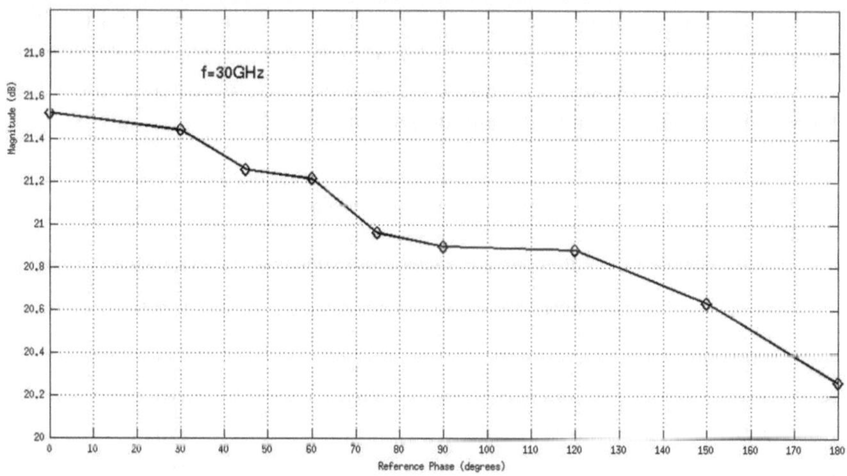

Fig. 2.21 Peak gain vs. reference phase: E_{φ} cut

c) *First Sidelobe Level*: As observed in Fig. 2.19, the first sidelobe level in the E_{φ} cut decreases by 4 dB from the 0° reference phase case with the lowest sidelobe being at the reference phase of 150°. The analysis of the E_{ϑ} cut shows a significant increase in the sidelobe level when the reference phase is 30°, but on careful observation this sidelobe is not in the same location as the other first lobes. The results are summarized in Table 2.5.

Table 2.5 Change in first sidelobe level and reference phase

F/D	First sidelobe decrease E_φ plane (dB)	Ref. phase E_φ plane (deg)	First sidelobe decrease E_ϑ plane (dB)	Ref. phase E_ϑ plane (deg)
0.237	2.81	180	8.44	180
0.35	2.19	180	12.7	180
0.5	2.73	120	15.26	180
1.0	5.94	120	3.72	180

d) *First Sidelobe Peak Location*: The location of the first sidelobe from boresight as the reference phase increases has shown that, other than the 30° reference phase in the E_ϑ cut, the location of the first sidelobe is generally constant. The E_φ cut has a more constant first sidelobe location than the E_ϑ cut.

The most significant observation is that the maximum sidelobe level decreased with increasing reference phase in the E_φ plane while the opposite occurred in the E_ϑ plane. As the reference phase increased, the maximum sidelobe level also increased in this plane. The largest maximum sidelobe decrease in the E_φ plane, out of all the reference phase cases, was 4.72 dB at a reference phase of 120° and $F/D = 1.0$. In the E_ϑ plane, the largest maximum sidelobe increase, out of all the reference phase cases, was about 4 dB at a reference phase of 180° and $F/D = 0.35$. In both planes, the largest increase/decrease occurred at the largest reference phase. These results are summarized in Table 2.6.

Table 2.6 Change in maximum sidelobe level and reference phase

F/D	Max. sidelobe decrease E_φ plane (dB)	Ref. phase E_φ plane (deg)	Max. sidelobe decrease E_ϑ plane (dB)	Ref. phase E_ϑ plane (deg)
0.237	2.81	180	2.66	180
0.35	2.19	180	3.93	180
0.5	2.73	120	3.23	180
1.0	4.72	120	2.41	180

e) *Cross-Polarization*: The important point is that, in both cases, the cross-polarization level for all reference phases is about 20 dB worse (*higher*) than the 0° reference phase case. This is clearly illustrated in Fig. 2.22, which shows the maximum cross-polarization levels in each plane.

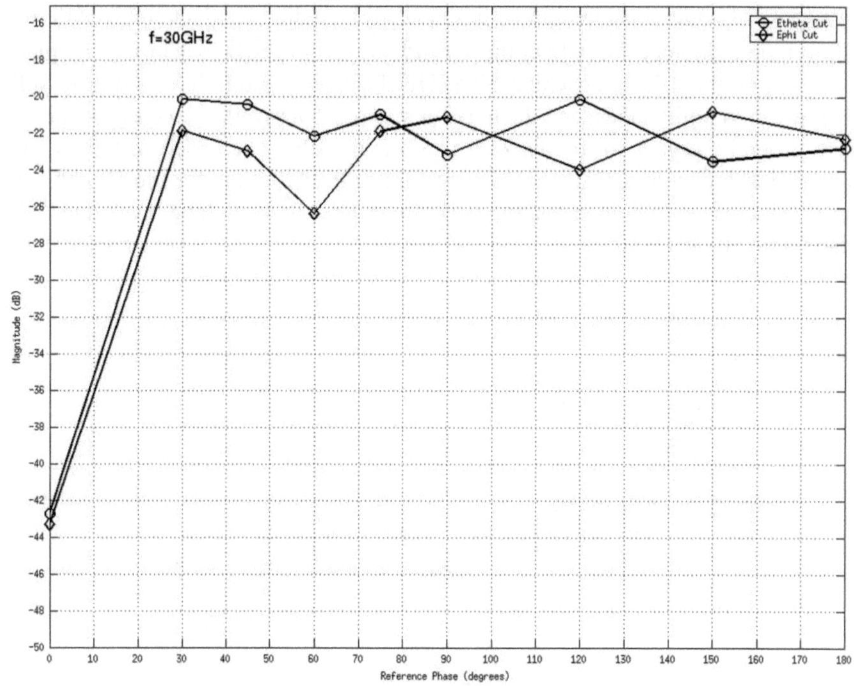

Fig. 2.22 Maximum cross-polarization vs. reference phase

f) *Half-Power (–3 dB) Beamwidth*: As the reference phase increased, the 3-dB beamwidth in all *F*/*D* cases was found to decrease by about 1°. The largest decrease in 3-dB beamwidth was 0.966 in the E_ϑ plane for *F*/*D* = 0.237 case and 0.853 in the E_ϑ plane for *F*/*D* = 0.5 case. A small portion of this decrease was a result of the slight change in zone diameter as the reference phase increased.

Thus, for the hexagonal FZP increasing reference phase from 0° to 180° causes about a 1.3 dB drop in peak gain. In the E_φ plane, there is about a 4 dB improvement in the level of the first sidelobe, with increasing reference phase from 0° to 180°. In the E_ϑ plane, increasing reference phase from 0° to 180° has a detrimental effect on the level of the first sidelobe. The cross-polarization level increases by about 20 dB in both planes as the reference phase changes from 0°.

2.2.5.4 Comparison of Circular and Hexagonal FZP

Figures 2.23 [21] illustrates a comparison of the co-polarized radiation patterns between the HFZPL and the CFZPL in the E_ϑ and E_φ planes for *F*/*D* = 0.237. In order to magnify the effect of the first sidelobe, the angle range is limited to the first 50° from boresight. From Fig. 2.23, the CFZPL shows a higher peak directivity of about 3 dB compared to the HFZPL. In the E_φ plane the CFZPL has a lower first sidelobe level than the HFZPL by about 8 dB, but in the E_ϑ plane there was only a 3.3 dB difference. The HFZPL was found to have a bigger 3-dB beamwidth in both planes, and the maximum cross-polarization levels (not shown) were generally higher for the HFZPL than the CFZPL. The observed trends with changing *F*/*D* for the CFZPL were very similar to the trends for the HFZPL. As the *F*/*D* increased in both cases, the peak directivity, first sidelobe level, and 3-dB beamwidth all decreased.

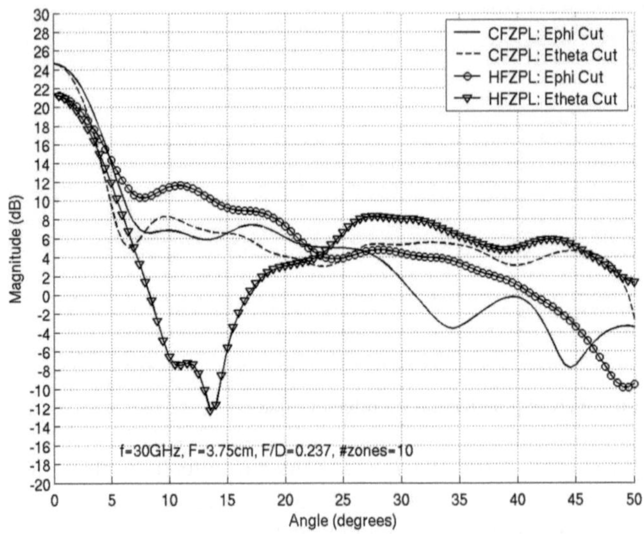

Fig. 2.23 Co-polarized radiation pattern comparison

Table 2.7 Summary of CFZPL and HFZPL simulation results

Circular FZPL			
F/*D*	Peak directivity (dB)	First sidelobe level (dB) E_φ / E_ϑ	–3-dB beamwidth (dB) E_φ / E_ϑ
0.237	24.65	–17.8 / –16.3	5.2 / 4.4
0.35	24.67	–16.6 / –16.6	4.6 / 4.2
0.5	23.63	–15.8 / –15.8	4.4 / 4.1
1.0	21.12	–14.4 / –13.8	3.8 / 3.6
Hexagonal FZPL			
0.237	21.31	–9.7 / –13.0	6.3 / 5.4
0.35	21.63	–7.7 / –14.1	5.7 / 5.3
0.5	21.11	–7.8 / –13.7	5.1 / 4.8
1.0	19.84	–8.2 / –13.5	4.2 / 4.1

The results vary depending on the *F*/*D* and are summarized in Table 2.7. As expected, the CFZPL has better antenna characteristics than the HFZPL, but the HFZPL is a fair approximation to the CFZPL performance. It should be noted that the drop in directivity with increasing *F*/*D*, which was common in both cases, was due to the feed not being optimal for the lens in the larger *F*/*D* cases. The feed was designed to be optimal for *F*/*D* = 0.237 case so that when the *F*/*D* was increased, the 10-dB beamwidth of the feed was much larger than that required for the 10-dB edge taper and therefore significant spillover resulted. In practice, higher gain horns would be used to feed the lenses with large *F*/*D* values.

The HFZPL and CFZPL antennas were different in the directivity and sidelobe results. The directivity of the HFZPL antenna decreased with increasing reference phase, whereas the directivity of the CFZPL antenna remained essentially constant. Also, although the sidelobe levels in the E_{φ} plane were similar to the CFZPL antenna, the result in the E_{ϑ} plane showed an increase instead of a decrease with increasing reference phase.

2.2.5.5 "Fractal" Evolution of the Square FZP

In comparison of rectangle FZP, it is possible to explore this area more deeply. Actually, one of the designs that could be compared and treated is the one that is shown in Fig. 2.24. It is like a "fractal" evolution of the square. The number of inverted corners can be growing to finally obtain the sawtooth version of a circle.

Therefore, whereas the CFZPL antenna showed a clear improvement in the sidelobe levels with minimal effect on directivity and cross-polarization as the reference phase increased, the HFZPL antenna does not. It can be concluded that the reference phase of 0° is the best for the performance of the single HFZPL antenna.

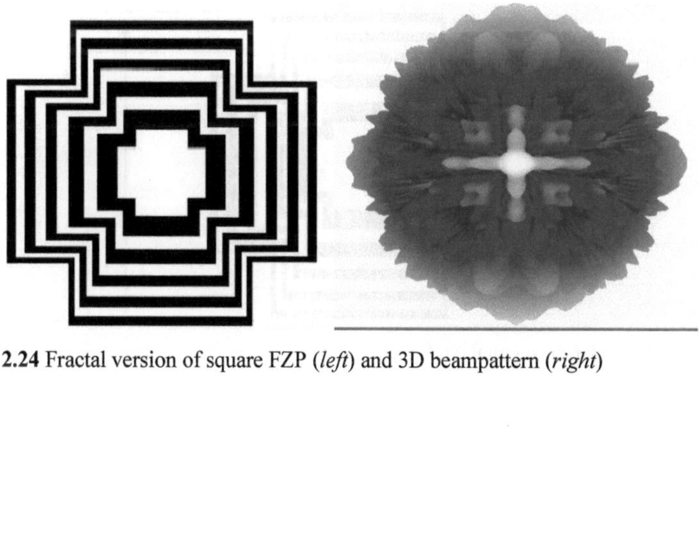

Fig. 2.24 Fractal version of square FZP (*left*) and 3D beampattern (*right*)

The dimensions of the circular, hexagonal, and fractal FZP are shown in Fig. 2.25. The comparative results of FDTD simulation of beampatterns for the circular, hexagonal, and "fractal" FZPs in E-plane are shown in Fig. 2.26.

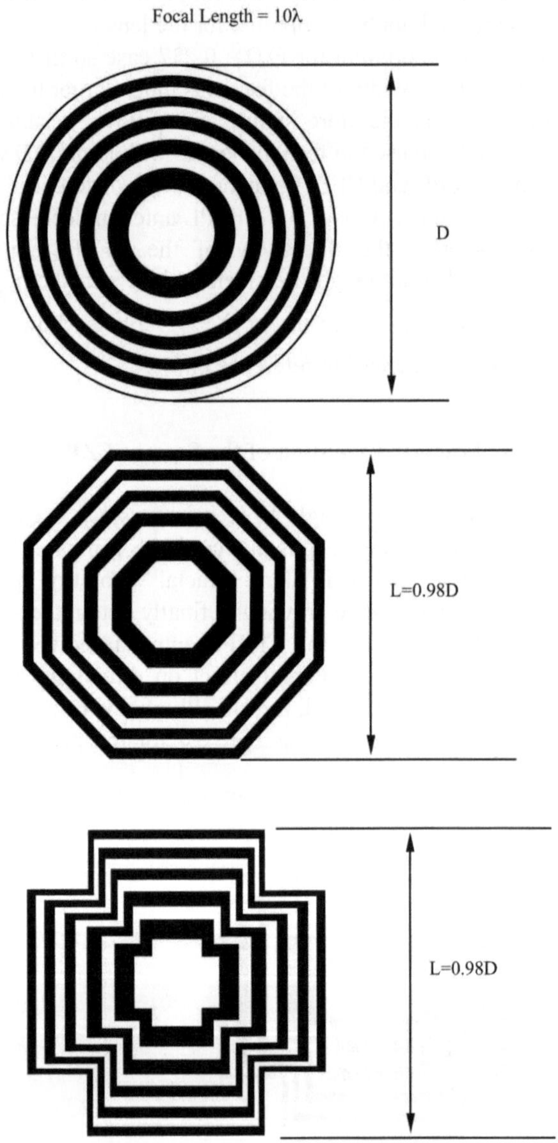

i	D_i	L_i
1	6.40	6.28
2	9.16	8.98
3	11.36	11.13
4	13.27	13.00
5	15.00	14.70
6	16.61	16.28
7	18.14	17.78
8	19.60	19.20
9	21.00	20.58
10	22.36	21.91
11	23.69	23.21

Fig. 2.25 Design of circular, hexagonal, and fractal FZP

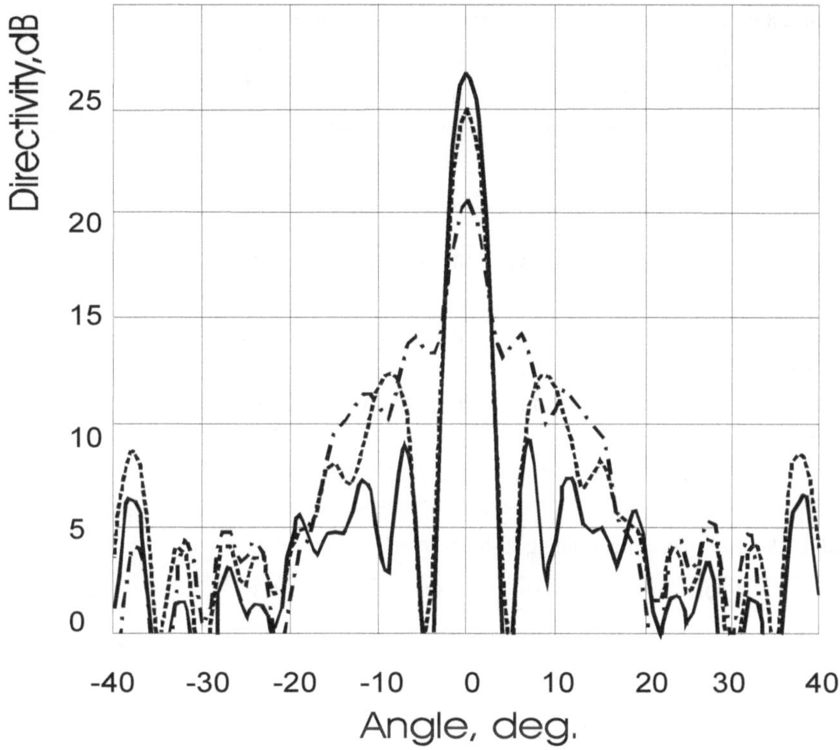

Fig. 2.26 FDTD simulation of beampatterns for the circular (*solid*), hexagonal (*dashed*), and "fractal" FZPs in E-plane

It appears that these modifications cause a significant drop in directivity compared to the circular FZP, but more effective than square FZP.

2.2.5.6 Antenna Having Star-Shaped Aperture

Let us pass to the results of calculations of WPT systems with transmitting antennas of star-shaped aperture [15]. Characteristic examples of the executed calculations for apertures with 16 and 32 beams are given in Tables 2.8 and 2.9. The normalized area of apertures at any value of F and N_{side} was kept constant and equal to one.

Table 2.8 Parameters of transmitting 16-beam star-shaped aperture

F	1.00	0.80	0.60	0.50	0.40	0.30
R_{min}	0.566	0.506	0.438	0.400	0.358	0.310
R_{max}	0.566	0.633	0.731	0.801	0.895	1.033
SL, dB	–17.6	–18.1	–20.3	–22.9	–28.4	–30.7
BW,%	100.0	98.3	92.0	86.9	80.0	71.0
PTR, $R = 1.0$	0.553	0.552	0.541	0.529	0.508	0.473
PTR, $R = 1.6$	0.784	0.778	0.751	0.723	0.768	0.609
PTR, $R = 2.0$	0.838	0.831	0.797	0.763	0.711	0.534

Table 2.9 Parameters of transmitting 32-beam star-shaped aperture

F	1.00	0.80	0.60	0.50	0.40	0.30
R_{min}	0.567	0.505	0.437	0.399	0.357	0.309
R_{max}	0.565	0.631	0.729	0.799	0.893	1.031
SL, dB	–17,6	–18.1	–20.2	–22.8	–28.3	–29.8
BW, %	100.0	98.2	92.3	87.0	80.0	71.1
PTR, $R = 1.0$	0.553	0.554	0.548	0.539	0.521	0.490
PTR, $R = 1.6$	0.783	0.781	0.761	0.737	0.635	0.630
PTR, $R = 2.0$	0.837	0.834	0.808	0.779	0.730	0.656

Analyses of calculated beampatterns have shown [15] that transition from round-shaped aperture to star-shaped aperture (while constant aperture area) is followed by increase of antenna dimensions and insignificant narrowing of main lobe beamwidth in angular electromagnetic field distribution in the vicinity of focusing point. These features of star-shaped apertures intuitive gave, for example, the basis to expect increase of power transfer ratio (PTR) in rectenna of constant size.

Unfortunately, due to the angularity of the aperture edge, some features of specific-distant sidelobes appear. Specific-distant sidelobes, despite of their rather low level, comprise big enough part from general radiated power. It means that these sidelobes appear rather wide and cover significant part of focal surface in the real angles region. Specific sidelobes can be superseded to the imaginary angles region only due to sharp increase of tooth number in star-shaped aperture. However such shapes of transmitting apertures are completely unsuitable for practical application, as a rule.

2.2.5.7 Diffraction on the FZPL with Irregular Boundaries

The expected effect is that the technique would be successful in reducing the sidelobe peaks. The average sidelobe level might not be changed very much.

In Fourier planes, scalar diffractive field in a far zone can be presented by Fourier-optics expressions [23]:

$$U(x,y,z)=\frac{\exp(ikz)}{i\lambda z}\exp\left[i\frac{k}{2z}\left(x^2+y^2\right)\right]\int U_i(x',y',0)\exp\left[-2\pi i(f_x^{'}+f_y^{'})\right]\mathrm{d}x'\mathrm{d}y'\cdot$$

In cylindrical system of coordinates (r,ϑ and ρ,φ are the cylindrical coordinates on the aperture and in a plane of observation):

$$\begin{cases}x=r\cos\vartheta\\ y=r\sin\vartheta\end{cases},\quad \begin{cases}f_x=\rho\cos\varphi\\ f_y=\rho\sin\varphi\end{cases}.$$

So

$$U(\rho,\varphi,z)=A\int_{-\pi}^{\pi}\int_{R(\vartheta)}U_i(r,\vartheta,0)\exp(-2\pi i\rho r\cos(\vartheta-\varphi))rdrd\vartheta,$$

where

$$A=\frac{\exp(ikz)}{i\lambda z}\exp\left(i\frac{k}{2z}(x^2+y^2)\right)\cdot$$

The limit of integrating $R(\vartheta)$ mathematically defines the inequalities of the aperture. For calculation of the specified integral, we shall present $R(\vartheta)$ in the manner of sum of the average radius a and functions $f(\vartheta)$ describing inequalities (see Fig. 2.27):

$$R(\vartheta)=a+f(\vartheta).$$

Fig. 2.27 Binary mask of the FZPL with irregular zones

Thus we have:

$$U(\rho,\phi,z) = A\int_{-\pi}^{\pi}\int_{0}^{a} U_i(r,\vartheta,0)\exp(-2\pi i\rho r\cos(\vartheta-\phi))r\mathrm{d}r\mathrm{d}\vartheta +$$

$$A\int_{-\pi}^{\pi}\int_{a}^{a+f(\vartheta)} U_i(r,\vartheta,0)\exp(-2\pi i\rho r\cos(\vartheta-\phi))r\mathrm{d}r\mathrm{d}\vartheta = A(U_1+U_2)\,.$$

It could be noted that the functional form $f(\vartheta)$ could be varied (fractals, random number modulation, etc.).

The integral from of the first composed represents the integrated form of record of Bessel function of the first type

$$U_1 = 2\pi a^2\frac{J_1(2\pi a\rho)}{2\pi a\rho}.$$

For calculation of the second composed (double integral from high-oscillating functions), we shall lead replacement of the new variables $\beta = r - a$, allowing to reduce it to the calculation of product of two integrals:

$$U_2 = a\int_{-\pi}^{\pi}\exp(-2\pi i\rho a\cos(\vartheta-\phi))\mathrm{d}\vartheta\int_{0}^{f(\vartheta)}\exp(-2\pi i\rho\beta\cos(\vartheta-\phi))\mathrm{d}\beta$$
$$+a\int_{-\pi}^{\pi}\exp(-2\pi ia\cos(\vartheta-\phi))\mathrm{d}\vartheta\int_{0}^{f(\vartheta)}\beta\exp(-2\pi i\rho\beta\cos(\vartheta-\phi))\mathrm{d}\beta\,.$$

Thus, expression for the calculation of a scalar field in a distant zone diffracted on an aperture in the screen in which edges $R(\vartheta)$ are described by any function is received.

For example, in Fig. 2.28 below the beampattern of the circular and offered FZPL are shown (the binary mask of the "irregular" FZPL is shown in Fig. 2.27 above).

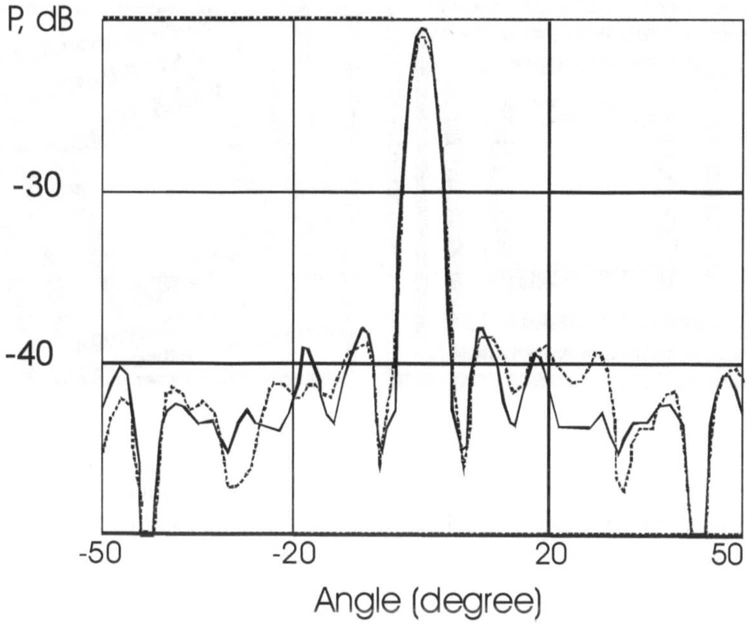

Fig. 2.28 Beampatterns of a circular FZPL (*solid line*) and FZPL with irregular zones (*dashed line*)

As followed from the investigations, the focusing is "robust" against zone boundary errors. The average sidelobe levels decrease and the optimization is possible.

2.3 Beampattern with Chiral Symmetry

2.3.1 Basic Principles of Zone Rotation

Let us consider a plane waves diffraction problem at the system of embedded one-by-one alternate (transparent/non-transparent) rectangular screens. As from above, in this case a filed intensity distribution at the distant zone in the first approximation is characterized by the function SINC.

The goal of the optimization is to eliminate the pronounced regular (cross-shaped) structure of the side maximums and to lower a level of the side maximums. To solve this problem Fresnel zones have been offered to be rotated with respect to each other by some angle which depends on the zone's number (Fig. 2.29) [16].

Fig. 2.29 Square FZP with classical and rotating zones

The zone plates diffraction is characterized by superposition of the diffracted waves on each embedded zone. Thus the field intensity distribution at the distant zone can be described by the following functional:

$$I(x,y) = \left| \sum_{n=1,3,5...}^{N} \begin{bmatrix} \sin c(\pi A a_n (x\cos\varphi_n + y\sin\varphi_n)) \bullet \\ \sin c(\pi A a_n (-x\sin\varphi_n + y\cos\varphi_n)) 2Aa_n^2 - \\ \sin c(\pi A a_{n-1} (x\cos\varphi_n + y\sin\varphi_n)) \bullet \\ \sin c(\pi A a_{n-1} (-x\sin\varphi_n + y\cos\varphi_n)) 2Aa_{n-1}^2 \end{bmatrix} \right|^2 \tag{2.12}$$

where φ_n is constrained by the following condition: a square bounds a_{2i} and a_{2i-1} must not intersect each other, that is:

$$-c(a_n) \le \varphi_n \le c(a_n), \text{ where } C(a_n) = \begin{cases} \dfrac{\pi}{4} : a_n > \sqrt{2} \cdot a_{n-1} \\ \dfrac{\pi}{4} - \arccos\left(\dfrac{a_n}{\sqrt{2} \cdot a_{n-1}}\right) \end{cases}, \tag{2.13}$$

$(a_0 = a_{-1} = 0)$, $\varphi_0 = 0$,

$A = const = \dfrac{2}{\lambda z}$; $a_i = \sqrt{i\lambda F + i^2\lambda^2 \cdot 0.25}, i = 1,2,3...$; $F, \lambda, z = const$;

In microwaves and for elements of micro-optics the number of zones as a rule is $N \sim 5 ... 11$, in millimeter-wave band $N \sim 20 ... 100$.

It can be shown that only first several zones can be rotated at a considerable angle. It follows from the zone bounds formula. Nevertheless, they are the first zones which make a maximum energy contribution.

Let us represent the problem mentioned before in terms of global optimization. Given a function $I_{\varphi_1,...,\varphi_n}(x, y): D \to [0,1]$ where $D = D_x \times D_y$ is a limited set. It is known that its maximum can be achieved at the point (0,0), $I_{\varphi_1,...,\varphi_n}(0,0) = 1$.

Let $W(\varphi_1,...,\varphi_n) = \max_{(x,y) \in L_I \setminus \{(0,0)\}} \{I_{\varphi_1,...,\varphi_n}(x, y)\}$ where L_I is a set of local optimas of the function $I_{\varphi_1,...,\varphi_n}(x, y)$. The function $W(\varphi_1,...,\varphi_n)$ is to be minimized using vector $\overline{\varphi} = (\varphi_1,...,\varphi_n)$ as a variable value, that is $W(\overline{\varphi}) \to_{\overline{\varphi} \in \Phi^n} \min$, where Φ^n is a set of allowed values limited by conditions (2.13). In other words, side maximum values of the function $I_{\varphi_1,...,\varphi_n}(x, y)$ are to be minimized by varying the zone's rotation angles.

As far as $W(\overline{\varphi})$ has a complex structure (e.g., when using a real physical model $W(\overline{\varphi})$ is a table function), it is reasonable to use numerical calculations aimed at answering the main question about $W(\overline{\varphi})$ optimization capability.

The function $W(\overline{\varphi})$ is complex enough so it is reasonable to use an approximation $W^{\varepsilon}(\overline{\varphi})$ instead of analyzing $W(\overline{\varphi})$. A value of $W^{\varepsilon}(\overline{\varphi})$ is obtained by multiple runs of random local search algorithm [17] (Table 2.10) from some number of points in D.

Table 2.10 Random local search algorithm

1. Select a starting solution $(x_0, y_0) \in D$. Record the current best-known solution by setting $(x^*, y^*) = (x_0, y_0)$ and $v^* = v_0 = I_{\varphi_1,\ldots,\varphi_n}(x_0, y_0)$.
2. Generate N_O solutions in a neighborhood $O((x^*, y^*), R)$ of the point (x^*, y^*) with a radius R according to uniform distribution on $O((x^*, y^*), R)$. Then choose a point (x', y') for which the following criteria is satisfied: $I_{\varphi_1,\ldots,\varphi_n}(x', y') > v^* + \delta$. If no such point is found, the algorithm stops.
Re-set $(x^*, y^*) = (x', y')$, $v^* = v'$ and return to step 2.

Let us choose a final number of points $D_\varepsilon \subset D$ such that for any $(x, y) \in D$ there is $(x', y') \in D$ and $\rho((x, y), (x', y')) < \varepsilon$, where ρ is Euclidean distance on D. Then random local search is started from every point in D_ε and it stops near one of the local optimas of the function $I_{\varphi_1,\ldots,\varphi_n}(x, y)$. In the set of local search results the maximal one which differs from (0,0) is selected. The resulting local optima is an approximate value of $W(\bar{\varphi})$.

Note that approximation accuracy is affected by the choice of ε while calculation time depends on local search parameters (the size of the neighborhood R and N_O – quantity of points to be examined). Exact estimations of approximation accuracy are likely to be found analytically but at this time we do not need them and empirical estimations can be used. For instance, after 100 starts of the approximation algorithm, a maximal deviation is not more than 10^{-4}. That is quite enough to make further calculations.

For $W^\varepsilon(\bar{\varphi})$ optimization on the set Φ^n the following heuristic global search algorithm is used [18] (Table 2.11).

Table 2.11 Global search algorithm

1. Set $i = 0$. Generate N_i points $\omega^{N_i} = \{\overline{\varphi}_1,..., \overline{\varphi}_{N_i}\}$ according to uniform distribution on Φ^n and compute values of $W^{\varepsilon}(\overline{\varphi})$ in these points. Let $\Omega^{N_i} = \{(\overline{\varphi}_1, W^{\varepsilon}(\varphi_1)), ..., (\overline{\varphi}_{N_i}, W^{\varepsilon}(\overline{\varphi}_{N_i}))\}$ be a set of the results obtained.

2. Select $k < N_i$ minimum values in Ω^{N_i}. Search perspective domain $\Phi^n_{N_i} \subseteq \Phi^n$ using classification obtained by one of the regression analysis algorithms (in this work LRP [9] is used).

3. Select next N_{i+1} points $\omega^{N_{i+1}}$ according to uniform distribution on $\Phi^n_{N_i} \subseteq \Phi^n$. Update Ω^{N_i} to $\Omega^{N_{i+1}}$.

4. If a termination criteria apply (such as limit on the total number of points to be examined), then the algorithm stops, else re-set $i := i + 1$ and return to Step 2.

At the Step 2 perspective domain is chosen. The problem of searching the domain $\Phi^n_{N_i} \subseteq \Phi^n$ is well known as classification problem [17]. Most algorithms solving this problem are based on a simple hypothesis that in average more perspective solutions are closer to the best found solutions than others. In this work LRP algorithm [18] is used. However, to increase algorithm efficiency other classification algorithms can be exploited (e.g., the nearest-neighbor classifier [19]).

Note that the global optimum is hardly to be found by the algorithm above [20]. This algorithm is aimed at searching solutions close enough to global optimum that allows us to answer the main question about appropriateness of studying the new type of Fresnel zones.

Furthermore, as follows from the algorithm description it can be applied to table functions. So a real antenna data can be used instead of mathematical model and the calculus accuracy can be noticeably increased by this way.

Some of function (2.12) optimization results are discussed below. The following parameters have been chosen for optimization: wave length $\lambda = 8$ mm, relative diameter D/λ=25, f-ratio F/λ = D/λ. For the function $W(\overline{\varphi})$ approximation on D a random local search with the following parameters is used: $N_R = 20$, $R = 1.5 \cdot \varepsilon$, $\delta = 10^{-4}$. Local search parameters could be selected according to a time/accuracy ration desirable (e.g. necessary parameters could be found by a simple enumeration). When using the parameters set from above the approximation error was in average less than 10^{-4}. The $W^{\varepsilon}(\overline{\varphi})$ optimization problem is solved by the global search algorithm with points limit (4000 function evaluations).

The changing of the maximum level of a side lobe in optimized Fresnel-zone plate relative to a number of full zones is shown in Fig. 2.30. Relative improvement (reduction of the maximum side lobe level) is 50.17%, 39.85%, and 39.33% for $n = 5$, 7, and 9 respectively.

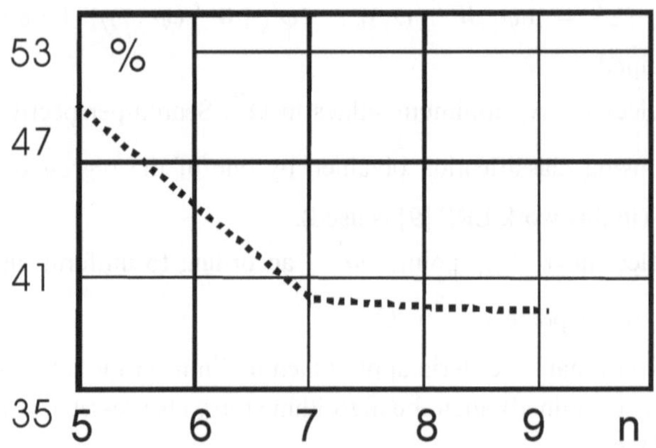

Fig. 2.30 Maximum sidelobe level versus number of full zones

One-dimensional section of intensity distribution along the line from (0,0) to a point of the maximum side lobe level is shown in Fig. 2.31.

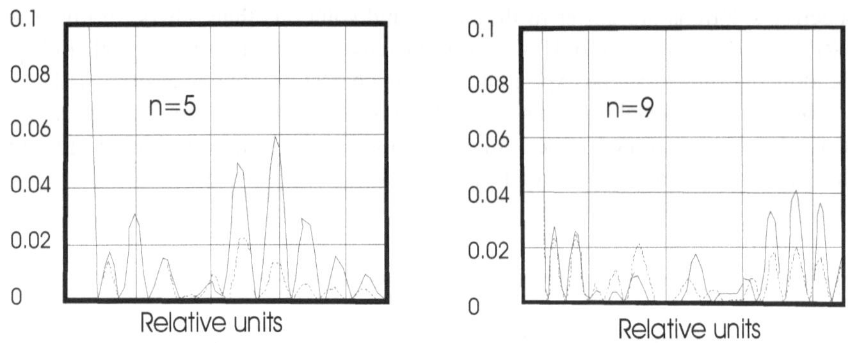

Fig. 2.31 *Solid line* – initial distribution, *dashed line* – optimization result

Increasing dimensions (number of full zones) makes the problem much more complex because additional dimensions greatly increase the size of the search space while reducing a classification algorithm's efficiency. To rotate only of a subset of several zones (e.g., from 1 to 3) partially solves this problem but finding the best number of full zones to be considered in optimization is left to further inquiries.

2.3.2 Square FZPs with Rotated Zones

FDTD simulations were carried out for the rotation in the same direction and for the rotation in alternating directions. The dimensions are shown in Fig. 2.32. Again, a vertical and a diagonal orientation were simulated. As examples the patterns are shown in Figs. 2.33 and 2.34 at 30 GHz, for the four cases. 3D beampatterns are shown in Figs. 2.35 and 2.36.

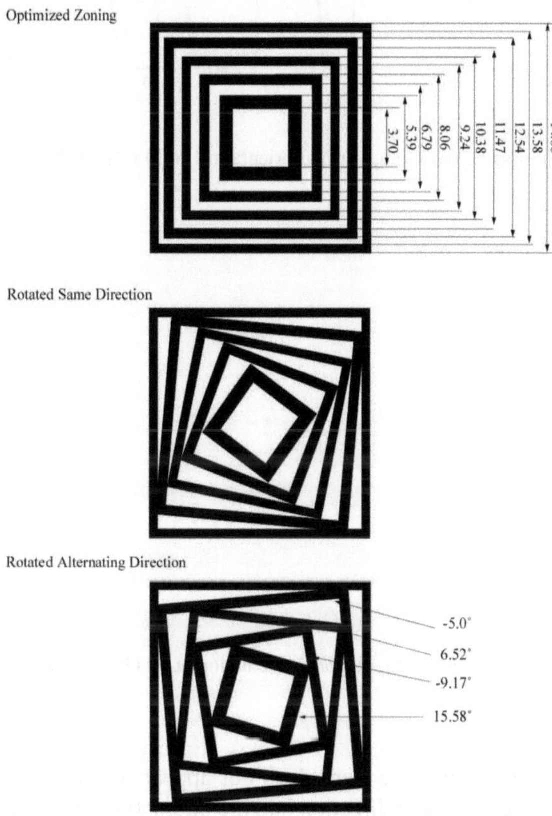

Fig. 2.32 Dimensions of square FZPs with rotating zones

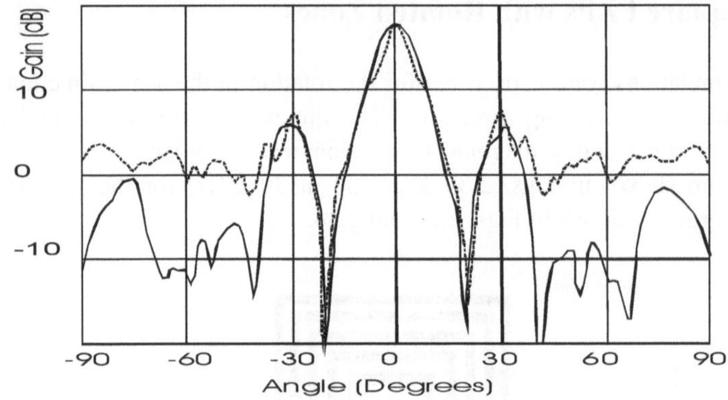

Fig. 2.33 Rotation in same direction, vertical orientation at 30 GHz: *solid line* $\varphi = 0$, *dashed line* $\varphi = 90$

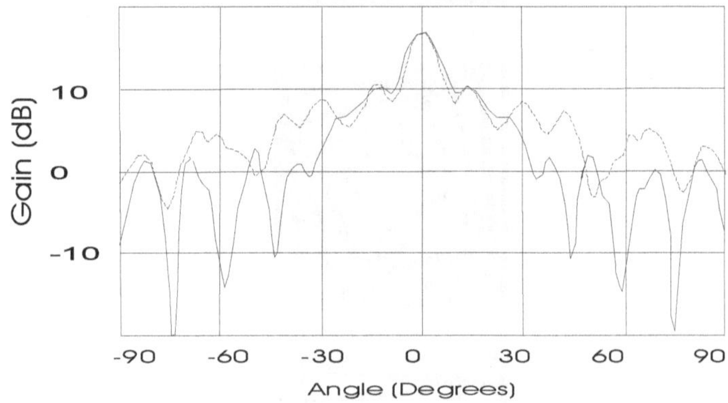

Fig. 2.34 Rotation in same direction, diagonal orientation at 30 GHz: *solid line* $\varphi = 0$, *dashed line* $\varphi = 90$

For the vertical orientation, there does not appear to be much difference between the rotation in the same direction and the alternating direction. However, for the diagonal orientation, the alternating direction has higher gain than the same direction.

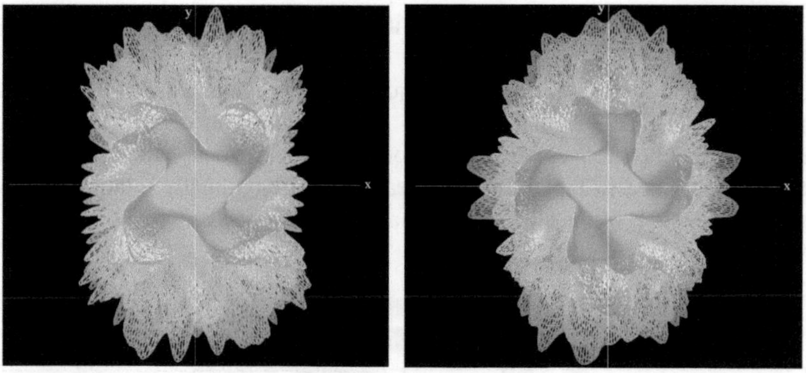

Fig. 2.35 3D beampattern of square FZPs with zone rotation in the same direction: vertical (*left*) and diagonal (*right*) orientation at 30 GHz. The chiral symmetry of sidelobes are clearly visible

Fig. 2.36 3D beampattern of square FZPs: effect of zone rotation. *Left* – vertical orientation without rotations, *center* – rotation on the same direction, *right* – rotation on the alternating direction

Figure 2.37 contains a plot of gain versus frequency for the vertical orientation. Gain versus frequency for the diagonal orientation show the same tendency.

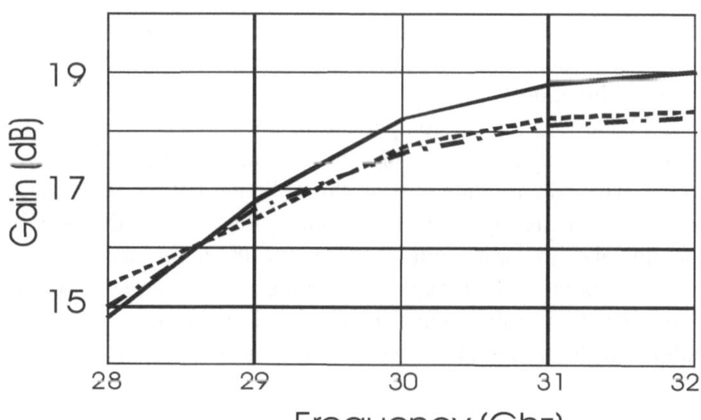

Fig. 2.37 The gain versus frequency for the vertical orientation: *solid line* – no rotation, *dashed line* – rotation on the same direction, *dotted line* – rotation on the alternating directions.

2.3.3 HFZPL Antenna with Rotated Zones

A concept introduced in [6] for a square FZPL antenna was investigated to improve the HFZPL antenna sidelobe performance. Since the hexagon is not circularly symmetric, it is possible to rotate the metal zones with respect to each other as illustrated in Fig. 2.38. The idea is that, by optimally rotating the zones, the sidelobe levels of the HFZPL antenna can be improved.

Fig. 2.38. Example HFZPL antenna with rotated zones

The lens parameters for these simulations were fixed with F/D=0.237 and reference phase = 0^0. The same open-ended WR28 waveguide was used to feed the lenses. It was placed at the focal point and again oriented such that the E field pointed in the *y*-direction.

Figure 2.39 shows a typical comparison of the HFZPL antenna radiation patterns in a three-dimensional view with and without the zones rotated with respect to each other. The primary sidelobes in the radiation pattern were blackened for ease of view. This figure highlights how rotating the zones of the HFZPL antenna caused the sidelobes to form a pin-wheel around the main beam. The rotation in this case was clockwise which explains why the sidelobes are curved in a clockwise fashion. Rotation in the counter-clockwise direction was found to create the same result but with the sidelobes curved in the opposite direction.

Fig. 2.39. 3D view of radiation patterns HFZPL antenna for different law of zone rotations

Alternating the zone rotation from clockwise to counter-clockwise was found to be ineffective and essentially voided the benefit of the rotation. Also, rotating only one zone, no matter which one, proved to have very little effect on the sidelobes.

The case that produced the best overall sidelobe level had all the zones rotated to their maximum extent such that they were not overlapping each other. In this case, the peak sidelobe level dropped by at least 2.6 dB and, more importantly, the location of the sidelobe was found to move further away from the main beam. The largest sidelobe movement was found to be about 15^0 but this could be further improved since an optimal arrangement of zones was not necessarily obtained. This effect is illustrated in Fig. 2.40 [21] where it can be seen that, in the E_φ plane, the absolute sidelobe level improved by 3.9 dB, and shifted away from boresight by about 8°. In all simulated cases, the peak directivity and 3 dB beamwidths remained essentially constant. The summary of beampattern parameters versus rotation is given in the Table 2.12.

Table 2.12 Peak directivity and peak relative sidelobe versus zone rotation

Rotation case	Peak directivity	Peak relative sidelobe
No Rotation	**21.43**	**–8.15**
A	21.43	–8.5
B	21.41	–8.97
C	21.21	–10.02
D	21.30	–10.23
E	**21.34**	**–10.44**

As it can be seen from the Table 2.4 the gain decreased less than –0.2 dB but relative sidelobe peak decreased up to –2 dB versus zone rotation optimization.

Thus as a means of potentially compensating for its sidelobe performance deficiencies, the zones of the HFZPL antenna were rotated in many different configurations. It was found that both the overall peak sidelobe level decreased and the sidelobes in the principal planes moved away from boresight. This ability to affect both the level and location of the sidelobe was important since it meant that the location of the sidelobe in the principal planes can be controlled by appropriately rotating the FZP lens zones.

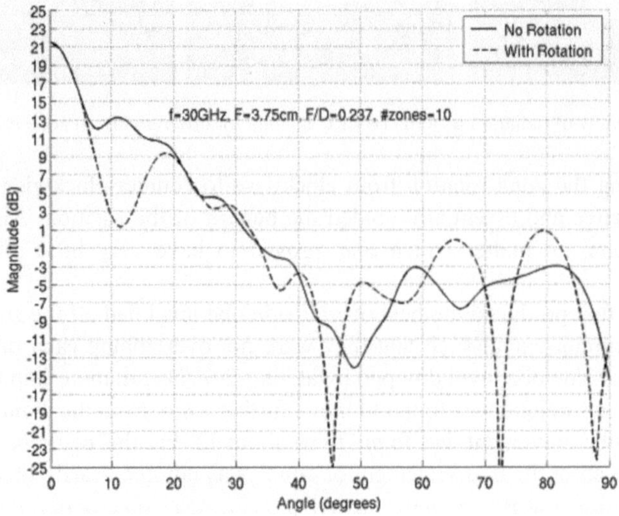

Fig. 2.40 Comparison with and without rotation HFZP antenna

2.3.4 FZP Antenna with Hexagonal-Cut Zones

An hexagonal FZPA, described above, was introduced as an alternative to the standard configuration of FZP. The primary advantage of the hexagonal geometry, as it was mentioned, was that it avoided overlap areas when the elements were used in an array. The secondary advantage was that the antenna's radiation patterns could be altered by rotating the zones with respect to each other since they were no longer circularly symmetric. This ability to control the location and level of the sidelobes is attractive for applications requiring specific sidelobe templates such as terrestrial wireless and certain satellite systems.

However, the overall gain and sidelobe level of the hexagonal FZPA were significantly worse than the standard configuration and thus would have limited practicality. Below, the hexagonal geometry is applied to the standard FZPA in a different manner [22]. Starting from the outermost zone, alternate circular zones will be trimmed, or cut, by a hexagon. In this way, the actual shape of the zones will be a mixture of the hexagonal and circular geometries. Figure 2.41 illustrates this proposed structure.

There are several benefits to this approach. First, because the trimmed zones continue to have partial circularity and only alternate zones are cut, the radiation patterns and gain will not be significantly degraded from the standard FZPA. Second, because the outer zone is trimmed to have hexagonal sides, the overlap problem is still avoided when the antenna is used as an element in an array. Finally, because some of the zones are not circularly symmetric, zone rotation is still possible for achieving the sidelobe control. This approach will therefore yield an improved FZPA which will have similar radiation performance to the conventional FZPA, but will have the added benefit of sidelobe control and will avoid overlap regions when used as an element in an array.

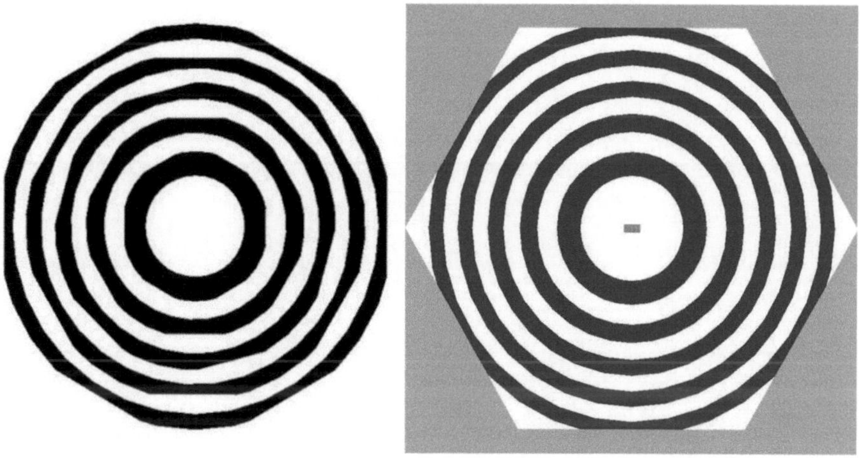

Fig. 2.41 FZP antenna with hexagonal-cut zones: all zones (*left*) and with outer ring cut in hexagon shape (*right*)

Compared with the standard version, the FZPA with alternating hex-cut zones had a decreased peak directivity of 0.32 dB and, in the *H*-plane, it had increased sidelobe levels by 0.73 dB. These numbers are a significant improvement over those of the FZPA with all hexagonal zones as it was mentioned above. Table 2.13 summarizes these results. The example of standard FZPA beampattern compared to FZPA with alternating hex-cut zones in *H*-plane is shown in the Fig.2.42.

Table 2.13 Summary of simulation results

FZPA configuration	Peak directivity (dB)	Gain (dBi)	Relative Max. SLB over all N (dB)
Standard circular zones	24.65	22.19	–16.30
Alternating hex-cut zones	24.33	21.92	–15.57
All hexagonal zones	21.52	19.00	–8.57

Simulation results also revealed that the size of the hexagon used to trim the circular zone had a significant impact on the peak directivity of the FZPA with alternating hex-cut zones. Removing too much of the circular zone would cause a decrease in the peak directivity. The optimal hexagon was found to yield the least amount of impact on the directivity.

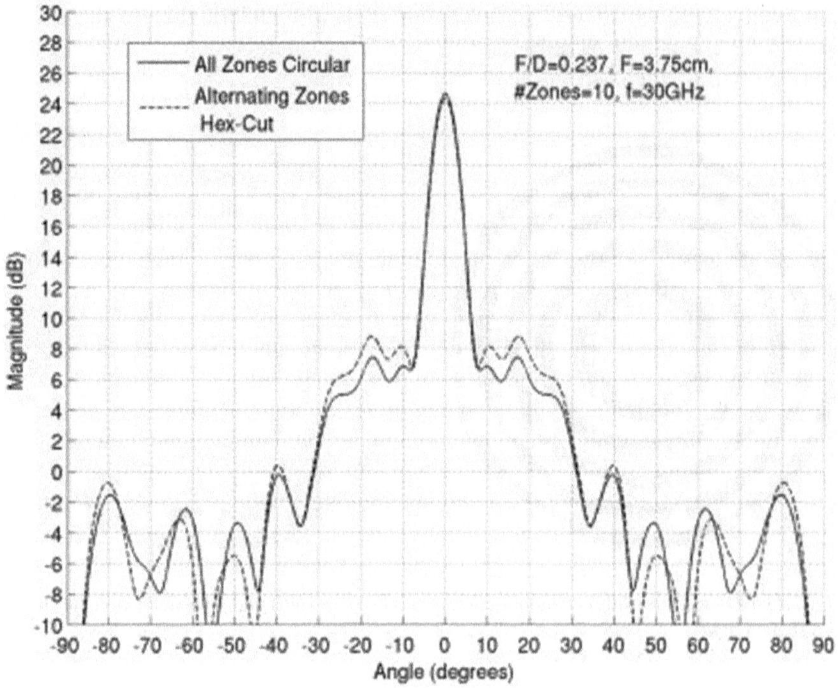

Fig. 2.42 Standard FZPA compared to FZPA with alternating hex-cut zones, *H*-plane

2.3.5 FZPA with Rotated Alternating Hex-Cut Zones

As it was mentioned, the main advantage of the new hex-cut FZPA configuration lies in the ability to control the sidelobe levels, something which cannot be done using conventional circular zones. In this study, many FZPA with alternating hex-cut (AHC) zones were simulated with varying configurations of rotated zones to observe the effect on the sidelobe levels and locations.

The maximum angle of zone rotation was 60° due to the symmetry of the structure. Table 2.14 summarizes a selection of these simulated results where the listings of rotation angles go from outer zone to inner zone. All rotations in the table are clockwise. It was found that rotations in the counter-clockwise direction produced the same results as those in the clockwise direction. Mixing the rotations from clockwise to counter-clockwise did not produce desirable results.

Several observations were made from the simulations:

- A maximum sidelobe level overall (better than the standard FZPA) by 0.12 dB.
- The peak directivity was unaffected by zone rotation.
- The relative maximum cross-polarization level increased from the no rotation case where it was effectively in the noise. However, the maximum cross-polarization was still better than 40 dB down from the co-polarized maximum. This change in cross-polarization level was expected since without the symmetry, the cross-polarization in the *H*-and *E*-planes would not cancel as it does in the symmetrical case.

Also, if the rotation was progressive from zone to zone, the sidelobes were found to form a pin-wheel while smearing along the direction of zone rotation. However, if the rotation was not progressive, the sidelobe smearing would not occur. Figure 2.43 illustrates this effect in the 3D far-field radiation pattern for three different cases, each cut at –18.5 dB from the normalized peak. Summary of measured results are shown in the Table 2.14.

Table 2.14 Summary of measured results

FZPA configuration	Peak gain (dBi)	Relative Max. SLB over all N (dB)
Standard Circular Zones	21.07	–15.28
No Rotation AHC Zones	20.76	–15.68
Rotated AHC	20.77	–16.00

These results show that the standard FZPA can have enhanced capability when alternate zones are hex-cut and then specifically rotated with respect to each other. Depending on the exact rotation configuration of the hex-cut zones, the sidelobes can be moved in nearly any direction and can also be smeared with one another. In this way, the designer can effectively place the sidelobes in a desired location. This could be quite useful for antennas in applications where a specific sidelobe template is required. Also, the enhanced capability comes with virtually the same radiation patterns as the standard FZPA. Rotating the hex-cut zones does not significantly degrade the radiation patterns or gain of the FZPA.

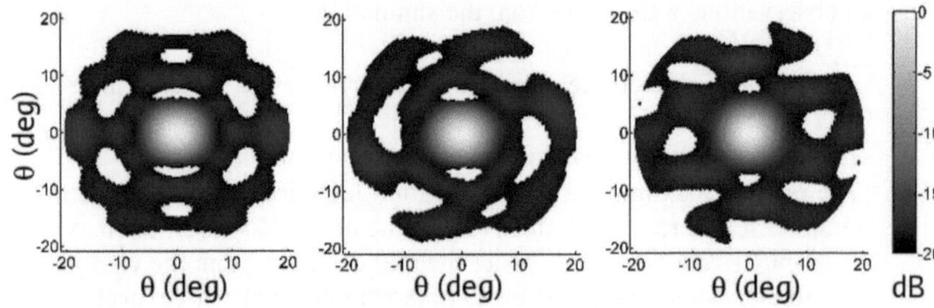

Fig. 2.43 Three-dimensional far-field radiation pattern for three different cases: (**a**) no rotation, (**b**) progressive rotation, (**c**) non-progressive rotation

2.3.6 CFZPL with Outer Ring Cut in Hexagon Shape

Far-field patterns of such FZP antenna (Fig. 2.41), in both planes, were very similar to the patterns of the original CFZPL simulations and shown in Fig. 2.44 as example.

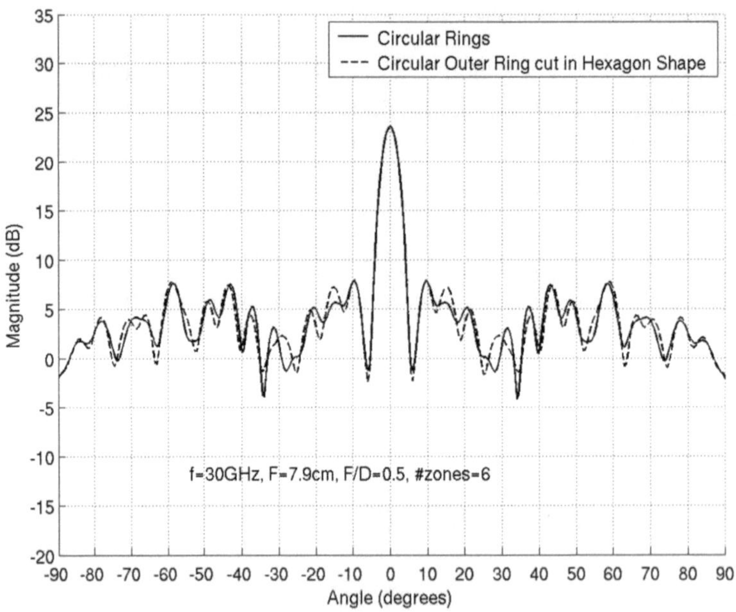

Fig. 2.44 Copol E_ϑ cut radiation pattern

References

1. S. M. Stout-Grandy, A. Petosa, I. V. Minin, O. V. Minin, J. Wight. A Systematic Study of Varying Reference Phase in the Design of Circular Fresnel Zone Plate Antennas. IEEE Transactions on Antennas and Propagation, Vol. 54, No. 12, December 2006, pp. 3629–3637.
2. O. V. Minin, I. V .Minin. *Diffractive optics of Millimetre Waves*. IOP Publisher, Boston, London, 2004.
3. P. A. Matthews, A. L. Cullen. A Study of the Field Distribution at an Axial Focus of a Square Microwave Lens. Monograph No. 186R, July 1956.
4. C. Polk. Optical Fresnel-Zone Gain of a Rectangular Aperture. IRE Transactions on Antennas and Propagation, Vol. AP-4, January 1956, pp. 65–69.
5. J. Gonzalez, J. Alda, B. Ilic, and G. D. Boreman. Infrared Antennas Coupled to Lithographic Fresnel Zone Plate Lenses. Applied Optics, Vol. 43, No. 33, 2004, pp. 6067–6073.
6. J. Ojeda-Castaneda and C. Gomez-Reino (editors). Thomson, B.J. (general editor), *Selected Papers on Zone Plates*, SPIE Opt. Engng Press, Washington, 1996.
7. L. J. Janicijevic. Diffraction Characteristics of Square Zone Plates. Journal of Optics (Paris), Vol. 13, No. 4, 1982, pp. 199–206.
8. I. V. Minin, O. V. Minin, A. Petosa, S. Thirakuone, Improved Zoning Rule for Designing Square Fresnel Zone Plate Lenses. Microwave and Optical Technology Letters. Vol. 49, 2007, pp. 276–278.
9. I. V. Minin, O. V. Minin. A Novel Type of Computer Optics Element. Abstracts of All-Union Seminar on "Methods and Technology of Processing Two-Dimensional Signals". Part II. Moscow: TsNIINTIKPK 1989. pp. 34–35 (in Russian). See also [2].
10. S. Ganci. Maggi-Rubinowicz transformation for phase apertures. Journal of the Optical Society of America A. Vol. 3, No. 12, December 1986, pp. 2094–2100.
11. M. Born, E. Wolf. Principles of Optics, 6th ed. Pergamon, Oxford, 1980.
12. I. Kallioniemi, J. Saarinen, K. Blomstedt, J. Turunen, Polygon approximation of the fringes of diffractive elements, Applied Optics Vol. 36, 1997, pp. 7217–7223.
13. Z. Chen, Y. Vladimirsky, M. Brown, O. Vladimirsky, F. Moore, F. Cerina, B. Lai, W. Yun, E. Gluskin. Design and Fabrication of Fresnel Zone Plates with Large Numbers of Zones. Journal of Vacuum Science & Technology B, Vol. 15 1997, pp. 2522-2527.
14. J. Alda, G. Boreman. Optimization of polygonal Fresnel zone plates. Microwave and Optical Technology Letters, Vol. 50, No. 2, 2008, pp. 536–541.
15. S. M. Stout-Grandy, A. Petosa, I. V. Minin, O. V. Minin, J. Wight. Hexagonal Fresnel Zone Plate Antenna. ANTEM 2006, Montreal, Canada, July 2006. pp. 511–514.
16. D. M. Sazonov, A. S. Shaposhnikov. Research of the WPT System with Transmitting Antenna Having Star-Shaped Aperture. Proc. 2002 12th Int. Crimean Conference "Microwave &Telecommunication Technology" (CriMiCo'2002). 9–13 September, Sevastopol, Crimea, Ukraine, pp. 271–274.
17. J. W. Goodman. Methods of Fourier Optics. – Stanford University, Stanford., 1968. 506p.
18. I. V. Minin, O. V. Minin, E. G. Danilov, G. S. Lbov. Parameters Optimization Algorithm of a new Type of Diffraction Optics Elements. Proc. 5th IEEE-Russia Conference MEMIA, Dec. 13–15, 2005, Novosibirsk, Russia, pp. 177–185.
19. L. A. Rastrigin. *Random Search.* Znanie, Minsk, 1979. 64p. (in Russian).
20. G. S. Lbov, N. G. Startseva. *Logical Solving Functions and Statistic Questions of Solving.* Novosibirsk, 1999. 212p. (in Russian).
21. R. O. Duda, P. E. Hart. *Pattern Recognition and Scene Analysis*, John Wiley, New York 1973. 512p.
22. I. V. Minin, O. V. Minin. Array of Fresnel Zone Plate Lens Antennas: Circular, Hexagonal with Chiral Symmetry and Hexagonal Boundary. Digest of the Joint 31st International Conference on Infrared and Millimeter Waves and 14th International Conference on Terahertz Electronics, September 18–22, 2006, Shanghai, China, p. 270.
23. S. M. Stout-Grandy, A. Petosa, I. V. Minin, O. V. Minin, J. Wight. Fresnel Zone Plate Antenna with Hexagonal-Cut Zones. Microwave and Optical Technology Letters, Vol. 50, No. 3, March 2008, pp. 672–676.

Chapter 3
FZP Lens Array

To reduce the profile of a lens antenna, one method is to design the lens with a lower value of *F/D* [1, 2]. But what is the minimal focal distance possible? Let us consider the focusing properties of FZP with sub-wavelength focus.

3.1 The Resolution Limit of FZPL with Small Values of *F/D* and Sub-wavelength Focus

For diffraction-limited optical and quasioptical systems, which are subject to paraxial approximation, the Rayleigh criterion gives the spatial resolution Δ for a circular lens of diameter D and focal length F as [2]:

$$\Delta = 1.22\lambda\frac{F}{D}. \tag{3.1}$$

The Rayleigh criterion defines the *xy* spatial resolution that can be achieved based on the wavelength of radiation that is used, the acceptance angle of the lens, and the index of refraction of the medium in which the radiation is propagating. From Eq. (3.1), it is seen that the spatial resolution decreases linearly with *F/D*. However, the paraxial approximation is generally agreed to be accurate when the value of $F/D \geq 0.5$ [3, 4], for which the spatial resolution is limited by $\Delta = 0.61\lambda$. It is well known that optical instruments are limited by diffraction, whose spatial resolution limit was originally determined by E. Abbe, to be approximately half a wavelength.

For optical applications, lenses are typically designed with values of F/D much larger than 1, due to the physically small values for the wavelength. These lenses thus have spatial resolutions much greater than a wavelength. At microwave and millimetre wave frequencies, since the wavelengths are much longer, as well as in microoptics, it is feasible to design lenses with values of $F/D < 0.5$. For these values, Eq. (3.1) is no longer an accurate estimate of the resolution. Although some studies were carried out on Fresnel lenses at microwave frequencies, only the impact of having small diameter lenses was investigated [5]. It is thus not clear what spatial resolution can be obtained for such designs. Moreover, for small values of *F*, the focal spot is in the reactive near field of the lens, and the effects of these reactive field components cannot be neglected, as is the case for larger values of F [6]. To investigate the behavior of such lenses, the focal fields of a phase-correcting Fresnel lens (antenna in receiving mode) were examined for several small values of F/D and with $F \leq 2\lambda$ [7, 8]. Until recently, the resolution of the FZPA with sub-wavelength focus remained unknown.

3.1.1 Some Details According Image Quality

As it is well known, a lens is a device that focuses electromagnetic (EM) energy. In transmit mode, a lens transforms the phase of a spherical wave originating from its focal point into a collimated beam or plane wave on the opposite side of the lens. In receive mode, it performs the opposite task of transforming an incoming plane wave into a spherical wave at the focal point. The lens accomplishes this transformation by introducing an appropriate phase change to the wavefront as it passes through the lens such that when the wavefront leaves the lens all parts of it will be in phase. In this way, a beam will be collimated.

Usually, the FZPL antenna is being operated as a transmission device. In this section, the opposite scenario is briefly discussed where the antenna is now in receive mode. This is the case where the receiver is located at the focal point of the lens and the plane waves passing through the lens are diffracted in such a way that they focus to the receiver. It is interesting to consider the focal fields in lens antenna designs because they allow verification of the correct operation of the lens in transmission mode [9]. Detailed description of the field intensity distribution at and near focus of FZP are discussed in [10]. For example, FDTD simulation has shown that (for a five-zone CFZPL antenna with $F/D = 0.237$ and $D = 15.8$ cm at 30 GHz), if the lens is focusing properly, there will be a peak at the desired focal point. This peak in this case occurs at 8.5 cm and the lens is located at 4.75 cm (based on the geometry of the structure in the simulation). This means that the focal point distance is 3.75 cm, which is correct based on the given F/D and the diameter.

The term "image quality" can be treated in different ways. In the general sense, the image quality is understood as a set of parameters characterizing the image of an extended object. In this case, besides system properties one should take into consideration such factors as the conditions of illumination, detection, and observation of the object and its structure. If it is necessary to describe the quality of the optical system in terms of aberration properties (as in optic waveband), one should examine the image of a point object (response to a pulse). In this case, one also needs to take account of system service conditions, the fashion of the image recording, for example. However, the influence of this and similar factors is not very strong and mostly reduces to the selection of criteria most suitable for the estimation of image quality.

A point image, just as an extended one, is described by a number of parameters that are facilitated by different numerical criteria used in estimating the image quality. These criteria should mostly satisfy two requirements: reliability and minimum labor content, i.e. a body of calculations needed to obtain this criterion. The labor content becomes very important when a system needs optimizing, i.e. when it is necessary to exhaust a great number of sets of system parameters to get the criterion of quality for each of them.

3.1.2 Main Well-Known Numerical Criteria of Point Image Evaluation

The complex amplitude distribution gives the fullest information about the point image. This distribution is derived by using the Fresnel–Kirchhoff integral on the base of the wavefront formed by an optical system at its exit pupil. Phase relations are, however, important in this distribution only when adjacent point images are superposed, i.e. in case of the extended source, which should be illuminated by highly coherent light as well. That is why the system spread function or optical transfer function is used in evaluating the image quality. The first function is a point image intensity distribution. It is known that, for rotationally symmetric optical systems and without aberrations, this distribution is the so-called Airy function (In 1834, Airy derived an expression for the intensity in the focal plane produced by a converging spherical wave limited in extent by a circular aperture [11].). Its center is located at the Gaussian image, and the distance from the center to the first minimum is called the Rayleigh resolution of the optical system [12]:

$$\delta_{\mathrm{R}} = \frac{0.61\lambda}{\sin(u)}, \tag{3.2}$$

where λ is the wavelength and u is the aperture angle of the system in image space, i.e. the angle of the system axis with a ray connecting an axial point at the image plane with an edge point of the exit pupil. So the Rayleigh criterion of resolution is: "*Two diffraction patterns with equal intensities may be said to be resolved when the central maximum of one pattern is not nearer than the position of the center of the first minimum of the neighboring pattern*". As a rule (in optical waveband), the quantity δ_R defines the limiting capabilities of an aberration-free (diffraction-limited) optical system. The Fourier image of a spread function is the optical transfer function (OTF). The module of the OTF, the modulation transfer function (MTF), provides, for example, the dependence of the contrast on the spatial frequency.

The following can be noted. As it was shown in [2], we can anticipate a priori that if, at a given wavelength λ, we have $\max(|ds|) \ll \lambda$ where ds is the optical path length difference for the rays connecting the radiation source to the surface of the sphere cantered at a point on the source image and having the radius equal to the rare segment of the objective lens, then waves will be focused in the same way as the aberration-free wavefront. In fact, we will be able to show that for a quasi-ideal wavefront it is sufficient to meet a less rigid condition,

$\max(|ds|) < \lambda/4$.

The data shown in [2] demonstrate that the real aberration wavefront differing from the spherical by not more than $\lambda/4$ generates the diffractional image of a point, which is practically indistinguishable from the ideal one.

Therefore, when the diffraction lens is optimized (especially a short-focus lens), it is advisable to use a criterion that minimizes the deviation of the wavefront behind the lens from the sphere whose radius and position of the center are specially adjusted (this can be done by minimizing the mean-square deviations of the front from spherical shape). But none of the numerical criteria has or can have full information about the optical system properties; therefore, depending on the features of an image, recording or observation conditions, it is necessary to use a criterion that is the most preferable under the circumstances.

The numerical criteria based on using the OTF are effective in evaluating the quality of teleobjectives, photographic objectives, and similar objectives, which usually have considerable residual aberrations (it should be noted for comparison that the Strehl intensity cannot be used here). In the image formed by these objectives for visual observation, details of different size are transferred with different contrast. Recording the image on a photographic emulsion is linear and retains, therefore there is difference in contrast. Hence, the knowledge of the amplitude and phase distortions with which different spatial frequencies are transferred permits to analyze the image quality most reliably.

A different approach is needed in evaluating the quality of an image formed by photolithographic objectives, the most perfect objectives at present. In this case, an image used for technological purposes is recorded in a photosensitive material with highly nonlinear properties, a photoresist, for example, which ensures the same contrast in recording details of any size, even the limiting size for a given objective (all errors are deviations of the size of details from a nominal value). It is then disadvantageous to use the criteria based on the OTF, which makes sense only in linear image recording. Moreover, photolithographic objectives with small residual aberrations form an image close to the diffraction-limited one, which also makes the image quality difficult to evaluate by means of the OTF. It is known that the OTF only slightly depends on the aberrations in the vicinity of limiting spatial frequencies [13], and, moreover, it is uninformative when aberrations are almost absent. More detailed comparison of different optical criteria may be found at [14].

It seems that it is best to evaluate the quality of photolithographic objectives by the level of energy in the diffraction pattern, by the part of the total energy, for example, that is concentrated within the Airy disk, the radius of which defines the resolution of the system (Eq. 3.2). It is reasonable to evaluate the quality of other types of objectives (the ones used in optical processing, for example) that also form an image close to the diffraction-limited one by means of either this or another criterion. Since optical systems based on diffractive elements considered in this book have small residual aberrations, we shall pay attention to the criterion evaluating quality by the energy concentration and to similar criteria.

Without aberrations, the relative energy inside the diffraction image disk of radius δ_R is equal to 0.84. Naturally, this is smaller in other cases. Let us set the least value of $E(\delta_R)$ at which an image can still be considered diffraction-limited by using the common image evaluation when a system has only a third-order spherical aberration. According to the Rayleigh criterion, an image is not practically different from the ideal one if the spherical aberration of an optical system does not exceed one fourth of a wavelength [13]. In this case, there is 0.73 of the diffraction spot total energy within the Airy disk. For systems with a low level of residual aberrations, we take the value $E(\delta_R) = 0.73$ as the limiting value of the energy concentration criterion. Though this value is rather arbitrary, it is well grounded and good practice from the authors' point of view. In this case, there are good grounds for extending the criterion limiting value obtained (or chosen) for one type of aberration to all other types, because it is quite clear that the same degree of energy concentration in the Airy disk provides practically the same conditions of recording an image (especially onto a nonlinear medium) independently of an aberration type. The fact that the criterion of energy concentration in the Airy disk is independent of the type of aberration makes this criterion more reliable than all other numerical criteria.

While calculating the relative energy $E(r_0)$ concentrated within the disk of the radius r_0 greater than the Rayleigh limit of the given optical system, one can evaluate the quality of the image and objectives with considerable residual aberrations. The expected system resolution at this point will be equal to the radius r_0, which provides the energy concentration $E(r_0)$=0.73.

3.1.3 Anomaly in a High-Numerical-Aperture Diffractive Focusing Lens

In almost well-known investigations of the focal distributions of a variety of focusing systems, it was determined that the spot size in the focal plane decreased monotonically. In the paper [15], the authors consider diffractive rather than refractive lenses with high numerical aperture (NA). They investigated the intensity distribution at and near the focus of a high-NA diffractive lens. And it was shown that diffractive lenses with high NAs behave differently than aplanatic lenses. Specifically, unlike the monotonic reduction in spot size with increasing NA that occurs with aplanatic lenses, in diffractive lenses there is an optimum NA value beyond which focusing degrades.

Using numerical calculation of Debye integral for the field at the focus, the authors of [15] calculated the intensity distribution at the focal plane as a function of NA, from which we deduced the spot size. For these calculations, two criteria were used. One was based on the first zero and the other on the encircled 84% of the energy (which corresponds to the first lobe of the Airy pattern).

As it was shown [15], for a low NA for which the scalar theory is valid, the spot size is the same for both aplanatic and diffractive lenses, regardless of which criterion is used, so the spot size obeys r=0.6 λ /NA. As the NA increases, the spot size of the aplanatic lens, based on the two criteria, as well as the spot size of the diffractive lens based on the first-zero criterion only, decreases monotonically, which can still be reasonably expected from the scalar theory. However, when the 84% encircled-energy criterion for the diffractive lens is used, the results reveal that there is a minimum spot size near NA$\approx$0.5, beyond which the spot size increases significantly. So it was shown an anomaly in which the energy spreads outward at the focus of a diffractive lens when the NA increases, thereby degrading the effective spot size. This behavior goes against intuition and common wisdom and could be taking into account the determination of energy criteria for image quality for diffractive focusing optics.

3.1.4 Results of FZPL Investigations with Sub-wavelength Focus

The binary phase-correcting Fresnel lenses used in this study consist of a set of annular dielectric rings whose radii were determined using the traditional Fresnel zoning rule for flat surfaces with the geometric optics approximation [2] and are given in Table 3.1. (It should be noted that the optimization of the Fresnel zone radii, i, for improved focusing was not the aim of this study.)

Table 3.1 Fresnel zone radii (in millimeters) for different focal length designs

i	$F = 2\lambda$	$F = \lambda$	$F = 0.5\lambda$
1	15.00	11.18	8.66
2	22.36	17.32	14.14
3	28.73	22.91	19.36
4	34.64	28.28	24.49
5	40.31	33.54	29.58
6	45.83	38.73	34.64
7	51.23	43.87	39.69
8	-	-	44.72
9	-	-	49.75

The thickness (t) of the rings is calculated based on the requirement of achieving a 180° path difference between the waves traveling through the dielectric rings and those traveling through air:

$$t = \frac{\lambda}{2\left(\sqrt{\varepsilon_r} - 1\right)}, \tag{3.3}$$

where ε_r is the dielectric constant of the rings. Three lenses were designed at 30 GHz using a dielectric constant of ε_r = 4, which from Eq. (3.3) result in a ring thickness of $t = \lambda/2$. The focal lengths for the lenses were chosen to be 2λ, 1λ, and 0.5λ. The resultant radii for these lenses are listed in Table 3.1. The diameters for the three lenses are somewhat different since an integer number of zones was used for each case; the values were chosen as close to D/λ = 10 as possible. The correspondent Fresnel numbers ($\pi D^2/\lambda F$) are between 150 and 600.

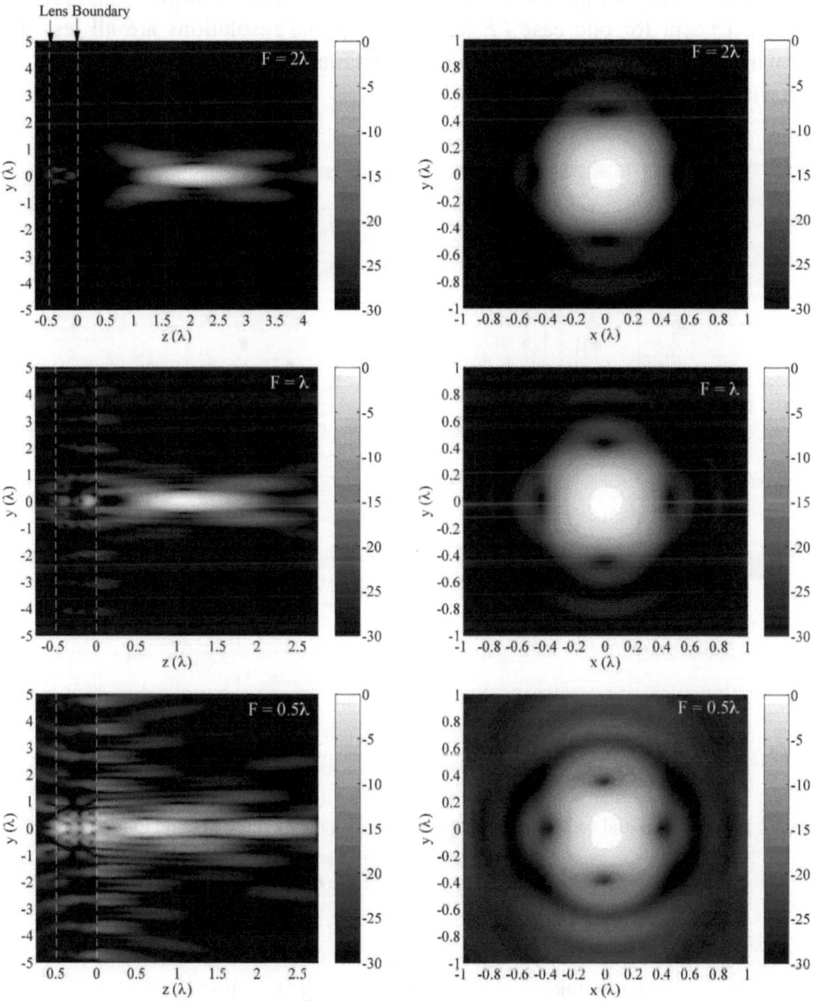

Fig. 3.1 Normalized power density in the yz plane for the three Fresnel lens designs (*left*) and normalized power density at the distance of maximum intensity (*right*)

The simulated focal fields in the yz plane (at $x = 0$) are shown in Fig. 3.1, where the focal spots are seen to the right of the lenses. For all three lenses, the maximum intensity occurs close to the designed focal length. This can be better observed in Fig. 3.2, which plots the power density (normalized to the peak value) along the focal axis ($x = y = 0$). The focal spots in the *xy* plane, taken at a distance from each lens where the focal fields had a maximum intensity, are shown in Fig. 3.1. The spatial resolution of the three lenses can be estimated from the location of the first nulls in the spot-beam pattern, the same rule as used by Rayleigh. Figure 3.3 shows the normalized magnitude for the spot beams, from which the resolution can be determined. The results are listed in Table 3.2. Since the spot beams were not perfectly axially symmetrical, the spatial resolution in both the *x*- and *y*-planes are listed. Except for one case ($F = 2\lambda$), the spatial resolutions are all less than 0.5λ, which is significantly finer than the spatial resolution achieved from the lenses with large values of *F*/*D*.

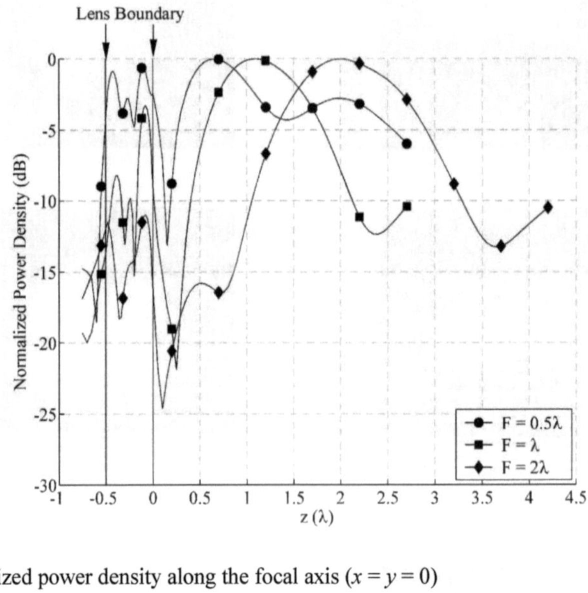

Fig. 3.2 Normalized power density along the focal axis ($x = y = 0$)

Table 3.2 Estimated spatial resolution and aperture angles of the three Fresnel lens designs

	$F = 2\lambda$	$F = \lambda$	$F = 0.5\lambda$
Aperture angle	137°	154°	169°
Resolution (*x*-plane)	0.52λ	0.45λ	0.40λ
Resolution (*y*-plane)	0.48λ	0.45λ	0.37λ

This preliminary investigation on the focusing properties of phase-correcting Fresnel lenses with values of *F/D* < 0.2 and with $F \leq 2\lambda$ has shown that a spatial resolution of less than 0.5λ is achievable. The "Abbe barrier" was thus completely broken by such lenses. As the *F/D* values decrease, so does the spot beam size, then a finer resolution is achieved. Although the spot beam decreases with decreasing *F/D*, the total amount of focusing power also decreases. This can be seen in Fig. 3.1, which shows the focal fields less well confined into a single spot as the *F/D* decreases. There is thus a tradeoff between refined resolution and focusing power, and the selection of the *F/D* will thus depend on the intended application.

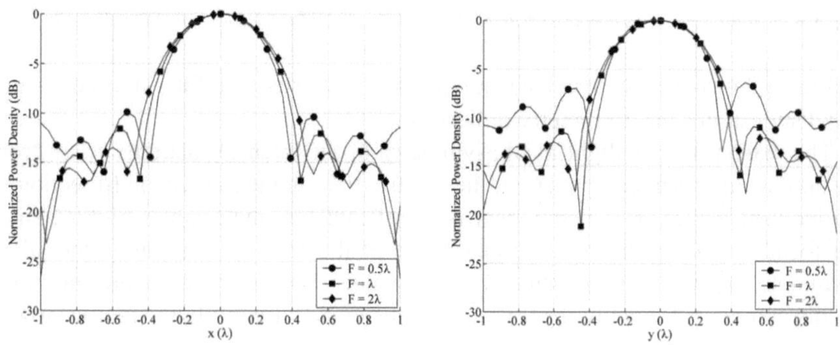

Fig. 3.3 Spot beams for the three Fresnel lenses

The spatial resolution was determined by determining the distance between the peak and the location of the first nulls in the spot-beam pattern. These results are summarized in Table 3.3, where it can be seen that the spatial resolution of the FZPA is also under 0.5λ and is close to that of the dielectric Fresnel lens, particularly along the *x*-axis.

Table 3.3 Comparison of spatial resolution between FZPA and dielectric Fresnel lens antenna

Antenna type ($F = 0.5\lambda$)	*X*-axis resolution (λ)	*Y*-axis resolution (λ) left side/right side
Fresnel Zone Plate	0.429	0.429 / 0.404
Dielectric Fresnel Lens (ε_r=4)	0.410	0.37 (average of left and right)

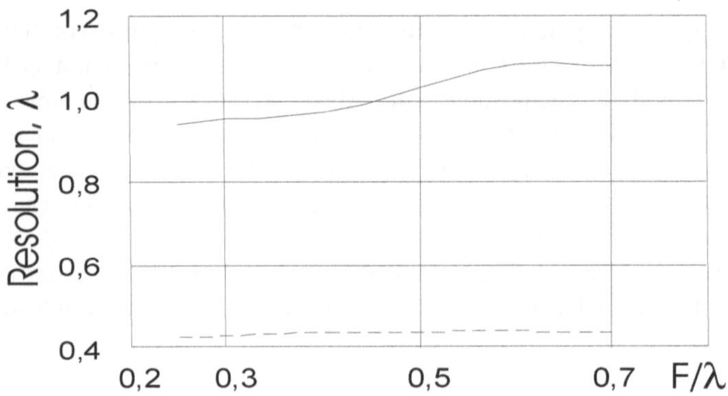

Fig. 3.4 Average resolving power versus F/λ for x-axis (*solid line*) and y-axis (*dashed line*)

Figure 3.4 shows the average resolving power versus F/λ for x-axis and y-axis and demonstrates the asymmetry of beam spot.

The reason for the slight axial asymmetry in the intensity patterns arises from the anti-symmetrical component of the electric field in the z-direction (direction of the incident wave). This component is significant for small values of *F/D*, since the focal spot is in the reactive near field of the lens, and it causes the slight asymmetry in the intensity patterns. As the values for *F/D* increase, the amplitude of this component quickly decreases and thus does not contribute significantly to the intensity, resulting in a symmetric pattern.

The ability of the FZPA to create a smaller spot size with increasing sub-wavelength focal spacing is related to the fact that the operation of the antenna depends more on the near fields than on the radiative field components used in the standard interference methods. Rayleigh and Abbe's limits apply to these standard interference methods but do not apply when the near fields are significant. This explains why sub-wavelength focusing can exceed the standard limits. Thus the minimal focal distance equal to half of wavelength is available.

3.2. Single-Zone FZPA Case

As it was mentioned, one of the disadvantage of the FZPA is its relatively large volume compared to planar arrays. Similar to shaped lenses and parabolic reflectors, the FZPA has a depth dimension that is dictated by its focal distance. Since this distance is typically several wavelengths long, the overall depth dimension of the FZPA is significant when compared to a planar array.

Planar arrays, however, have their own problems at Ka-band. The microstrip feed networks are much less efficient at these frequencies and there are many elements to feed. This is another area where the FZPA is advantageous. A low-profile FZPA array would also likely involve a microstrip feed network, but there would be less elements and therefore less loss. Achieving a small focal distance FZPA would therefore make an attractive low-profile antenna, which can be a viable alternative to planar arrays and be potentially used in spatial power combining applications.

The modeling in this part of the study involved only a single metal zone as shown in Fig. 3.5 (left). In this part, the focal distance of the FZPA was decreased from 3.75λ to 0.25λ in increments of 0.5λ. In Fig. 3.5 (left), the ground plane was 10 × 10λ, which represented the minimum size of ground plane required for the microstrip patch antenna.

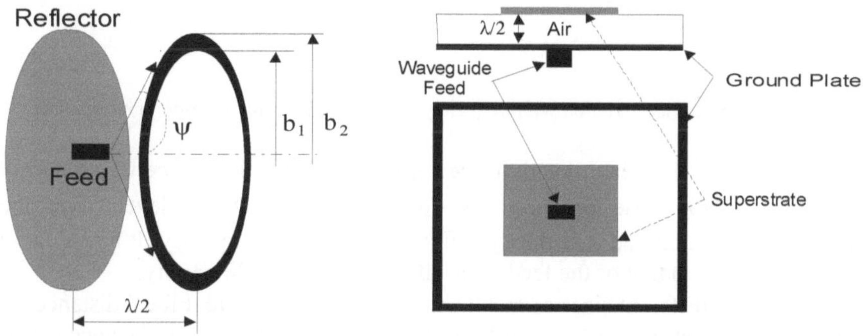

Fig. 3.5 Single-zone FZP (*left*) and $\lambda/4$ superstrate antenna Spaced $\lambda/2$ above a ground plate (superstrate $\varepsilon_r = 10$, $h = \lambda/4$)

For each focal distance, the radii of the metal zone were recomputed using Eq. (1.17). This equation is valid only for large focal distances ($F > 2.8D$). It depends on the principle of ray-tracing, which requires that the rays make relative small angles (<10°) with the focal axis so that they can be paraxial. In the case of small focal distances ($F < 2.8D$), Eq. (1.17) can be used only as a starting point with the understanding that the resulting radii would not be accurate [2].

The ground plane around the aperture of the waveguide was added in order to approximate the performance of the FZPA when fed with a microstrip patch antenna. The microstrip patch is the preferred feed for this structure due to the resulting attractive compact package where the feed, a layer of foam, and the lens can be all sandwiched together. However, in order to focus exclusively on the performance of the FZPA without the added interaction from the patch substrate, it was decided to use the waveguide feed with a ground plane instead for this study.

Figure 3.6 illustrates the simulated peak directivity variation with frequency for each focal distance with and without the ground plane, respectively [16]. For the case with the ground plane, the peak directivity curve generally decreased with increasing frequency for each focal distance until it reached about 30 GHz, then it would increase again, reaching a maximum at a frequency higher than 30 GHz. These curves were quite different compared to the same structures without the ground plane at the aperture of the waveguide. Without the ground plane, the peak directivity maintained nearly a constant level with varying frequency. This indicated that the addition of the ground plane caused the peak directivity to be fairly frequency-sensitive and less broadband.

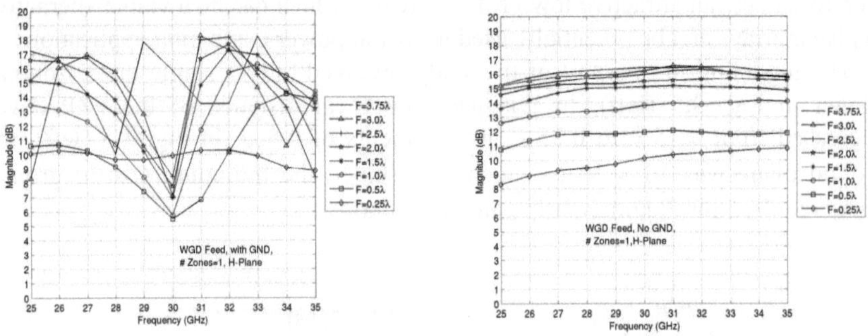

Fig. 3.6 Peak directivity variation with frequency: with (*left*) and without (*right*) ground plate

As expected, the peak directivity decreased as the focal distance was reduced. The $F = 0.25\,\lambda$ case produced the lowest peak directivity levels over the frequency band and had a distinctly different curve with frequency. It had a curve that was similar to that of the feed by itself, but with more directivity.

Due to the different sizes of metal zone radii for the different focal distances, a direct comparison was only possible for the 3.75 and 1.25 λ focal distance cases, where the aperture diameter was kept constant at about 6 λ by having two metal zones in the 1.25 λ case and one metal zone in 3.75 λ case. The smaller focal distance yielded a broader band response and a higher directivity at 30 GHz. The frequency sensitivity of the $F = 3.75\,\lambda$ case is caused by the larger separation between the lens and the ground plane. This sensitivity is not present when the ground plane is removed.

Figure 3.7 illustrates the H-plane radiation patterns of the FZPA with various focal distances at 30 GHz, which shows how only the 3.75 λ focal distance was well focused at this frequency. The shift in frequency confirms that the zone radii were not the proper size to obtain the peak at the designed frequency of 30 GHz. This was expected based on Eq. (1.17) not being valid for small focal distances due to paraxial approximation.

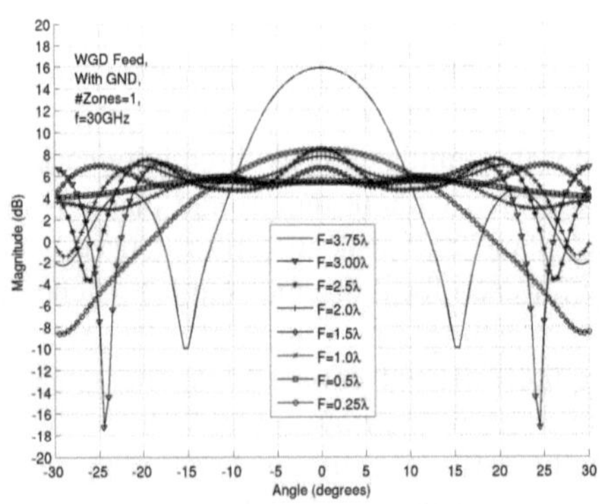

Fig. 3.7 Radiation patterns of the FZPA with various focal distances, H-plane

Based on the simulated peak directivity values and the physical size of the single-zone FZPA apertures, the effective area and antenna aperture efficiency were calculated. Some general statements can be made about the results. The aperture efficiencies FZPs with ground plate are generally higher than those without ground plate, confirming that the ground plane improves the antenna performance. The difference in aperture efficiency between having the ground plane and not having the ground plane for the $F = 1.5\lambda$ and 1.0λ cases is nearly 10%, which indicates that these two focal distances are more optimal than the others for this FZPA structure. In the $F = 0.25\lambda$ case, the aperture efficiencies are nearly the same with or without the ground plane. This was not surprising since the lens was unable to provide focusing regardless of the presence of the ground plane and so the actual radiation pattern is a version of the feed's radiation pattern.

This was the more desirable of the two approaches since it did not involve increasing the thickness of the overall antenna. In order to perform the zone size adjustment, the $F = 0.5\lambda$ case was selected and the zone radii that were computed based on Eq. (1.17) for this case were modified to be bigger by 5, 10, 20, 25, and 30%. Figure 3.8 illustrates the results of these simulations.

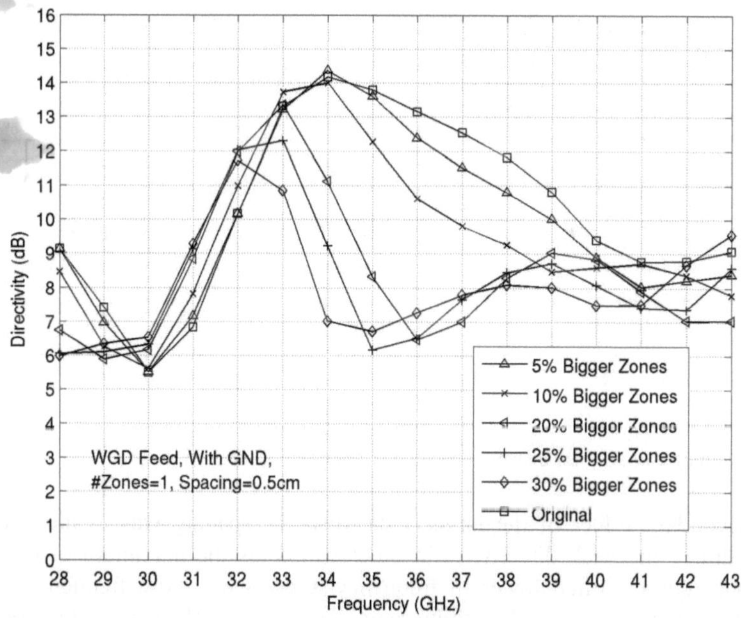

Fig. 3.8 Comparison of zone size variation for $F = 0.5\lambda$

As expected, increasing the zone size produced the desired effect of shifting the frequency of the peak directivity lower. Unfortunately, the bandwidth and the peak directivity dropped progressively with each increase in zone size. The 3-dB bandwidth dropped substantially and the peak directivity dropped by over 2 dB when the zone size increased to 30%.

3.2.1 Focal Distance Spacing Adjustment

This approach involved varying the spacing between the waveguide feed and lens of the FZPA without changing the radii of the zones. The zone radii were computed using Eq. (1.17) and were based on the focal distance of 0.5 λ. The zone size was kept constant for each simulation. The focal distance spacing was increased from 0.5 to 0.7 λ in steps of 0.05 λ. Figure 3.9 shows the resulting effect on the peak directivity versus frequency graph. As the spacing increased, the desired lowering in frequency of the maximum peak directivity was observed.

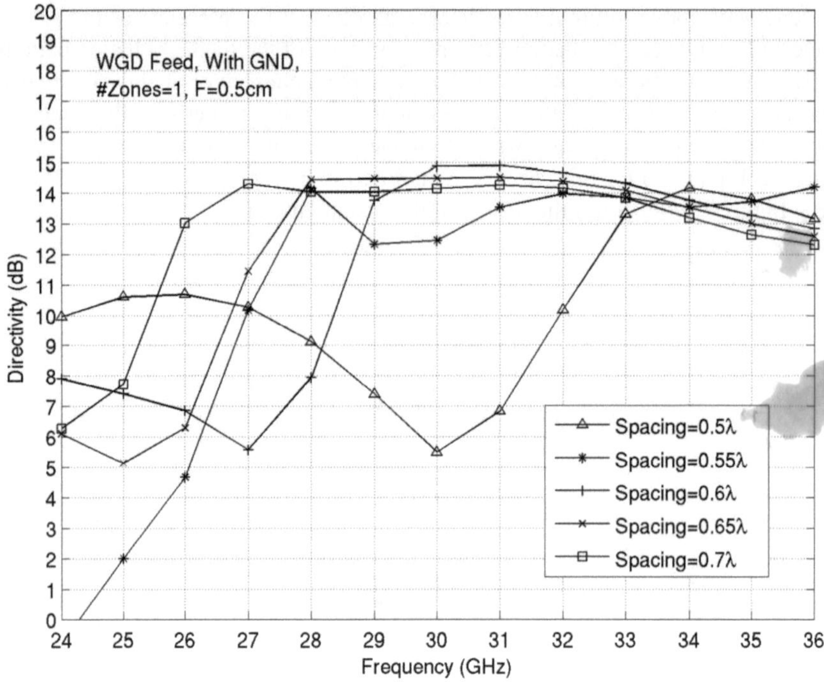

Fig. 3.9 Comparison of focal distance spacing variations for $F = 0.5\,\lambda$

A secondary effect of the larger focal distance spacing was to increase the bandwidth of the peak directivity. In other words, as the maximum of the peak directivity curve shifted lower in frequency, there were more frequencies over which the level of the peak directivity remained constant. The case that yielded the largest peak directivity 3-dB bandwidth was the focal distance spacing of 0.7 λ while the case that yielded the highest peak directivity was a focal distance spacing of 0.6 λ. The 0.7 λ focal distance spacing case was preferred though due to the improved bandwidth. Table 3.4 shows a comparison between 0.5 λ and 0.7 λ focal distance spacing at 30 GHz in terms of aperture efficiency.

Table 3.4 Comparison between 0.7 λ and 0.5 λ focal distance spacing at 30 GHz

Focal distance spacing (cm)	Directivity (dB)	Frequency of directivity (GHz)	Wavelength λ (cm)	Effective area (cm^2)	Physical area (cm^2)	Aperture efficiency (%)
0.5	5.49	30	1.00	4.04	6.28	4.48
0.7	14.14	30	1.00	2.06	6.28	32.87

By increasing the focal distance spacing to 0.7 λ, there was an improvement in aperture efficiency of 28%. Also it was found that the FZPA could not focus the fields at distances below 0.5 λ.

3.2.2 Single-Zone FZPA Compared with Superstrate Antenna

It was interesting to compare the single-zone FZPA, with ground plane surrounding the waveguide aperture, to an antenna based on a quarter-wave thick superstrate, spaced a half-wave above the same feed with the ground plane [17] – see Fig. 3.5 (right).

A superstrate in this configuration behaves as a leaky-wave antenna and is able to efficiently collimate a beam. In this comparison, only the 0.5 λ focal distance FZPA was considered since it was the only one that corresponded to a half-wave spacing between the feed and the superstrate. Also, the dielectric constant of the superstrate was chosen to be 10 and the size of the superstrate was set to be 25 × 25 mm square. This size represented approximately the same area as the outer ring of the single metal zone in the FZPA. Many other superstrate sizes were modeled, but this size yielded the maximum directivity at 30 GHz.

It is known that in the first-order scalar approximation the total field radiation pattern of antenna systems consisting of antenna and ring director may be given as follows [18]:

$$F(\vartheta) \approx F_a(\vartheta) + (F_a(\psi_1) + F_a(\psi_2))F_{\text{dir}}(\vartheta)(1-\cos(\vartheta))\exp(-i\beta d_1),$$

where (ψ_1, ψ_2) are the angles correspondent to the radii of a ring director (b_1, b_2) – see Fig. 3.5, $F_a(\vartheta)$ – the normalized primary antenna pattern, $F_{\text{dir}}(\vartheta)$ – the normalized diffraction pattern of the ring director, given by:

$$F_{\text{dir}}(\vartheta) \approx \frac{7.59[3.47J_0(u_1) + 4.12J_0(u_2)] + iu_1[3.47J_1(u_1) + 4.12J_1(u_2)]}{7.59^2 - u_1^2}.$$

Here $u_1 = \frac{2\pi}{\lambda} b_1 \sin\vartheta$ and $u_2 = \frac{2\pi}{\lambda} b_2 \sin\vartheta$, (J_0, J_1) – are the Bessel functions of zero and first-order, respectively. Note that for $u_1 \rightarrow 7.59$, $F_{\text{dir}}(\vartheta)$ is equal to $0.202\exp(i121(\text{deg}))$ [18].

Figure 3.10 shows a comparison at 30 GHz of the H-plane radiation patterns between the single-zone FZPA with $F = 0.5\lambda$ and the superstrate. Because the maximum directivity of the FZPA occurred at 34 GHz instead of 30 GHz, the difference in maximum directivity between the two is more than 10 dB at 30 GHz.

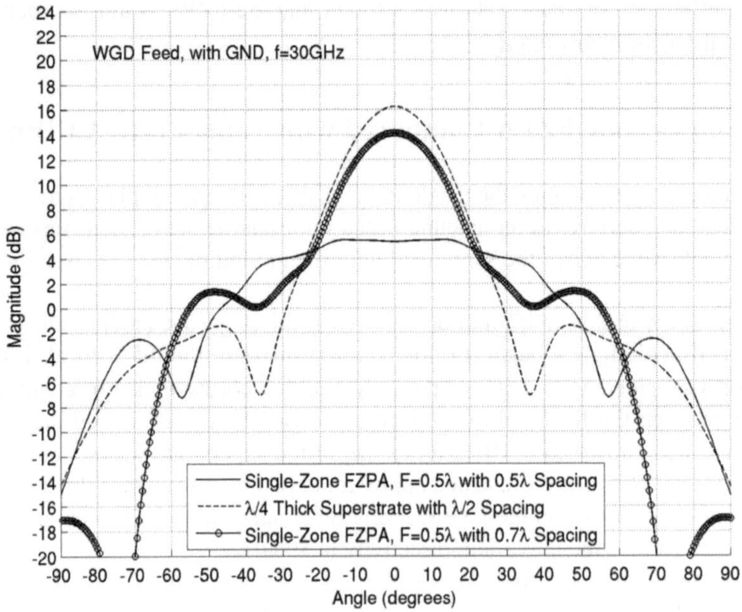

Fig. 3.10 Comparison between FZPA with 0.5λ and 0.7λ spacing and superstrate at 30 GHz (H-Plane)

However, by increasing the focal distance spacing of the FZPA from 0.5 to 0.7λ, while maintaining the same zone size, the frequency of the maximum peak directivity became closer to 30 GHz and the difference in peak directivity between the two antennas improved to about 3 dB. Figure 3.10 also shows this result. Despite the improvement in the FZPA compared to the superstrate structure, the superstrate still outperformed the FZPA in terms of peak directivity. However, the FZPA with ground plane around the waveguide aperture had a larger peak directivity bandwidth. Figure 3.11 shows the peak directivity versus frequency of each structure. The 3-dB bandwidth of the FZPA is 6.8 GHz whereas the 3-dB bandwidth of the superstrate is only 2.85 GHz. The other advantages of the FZPA over the superstrate are the practicality of the structure at lower frequencies and the cost-effectiveness of the overall package. The superstrate structure can be very costly due to the expensive microwave material required to obtain a high dielectric constant and low loss tangent. Also, at lower frequencies, the superstrate is not practical due to the large thickness requirements that also serve to further increase the cost.

So the FZPA was compared to a leaky-wave antenna with a $\lambda/4$ superstrate spaced $\lambda/2$ above a ground plane. This comparison revealed that although the superstrate antenna produced a higher peak directivity, the FZPA had a larger directivity bandwidth and would be far more cost-effective, especially at lower frequencies.

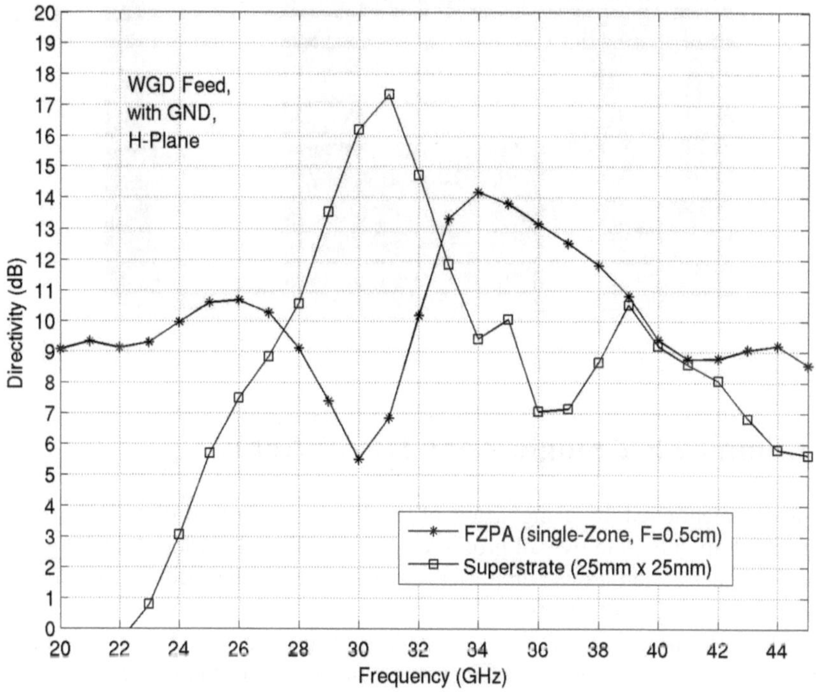

Fig. 3.11 Bandwidth comparison between the FZPA and the superstrate

In order to verify the simulated results, several antennas were selected for fabrication with single-zone FZPAs. The antennas were made using the same zone radii as in the simulations and were etched on a 0.127-mm FR4 material. They were each attached with tape to a piece of foam cut to be the thickness of the focal distance. A steel ground plane was fabricated to be the same dimensions as in the simulations with a rectangular hole in the center for the open-ended WR28 waveguide feed. The foam holding the antenna was attached to the ground plane. The entire structure was then situated on a custom-made test fixture. Figure 3.12 shows the test setup in the far-field anechoic chamber.

The measured peak gain was lower than the simulated gain by about 0.7 dB. The minor differences that were observed in the gain and sidelobe areas are likely due to the fact that the test fixture was not included in the simulations and the exact dielectric constant of the foam and FR4 was unknown.

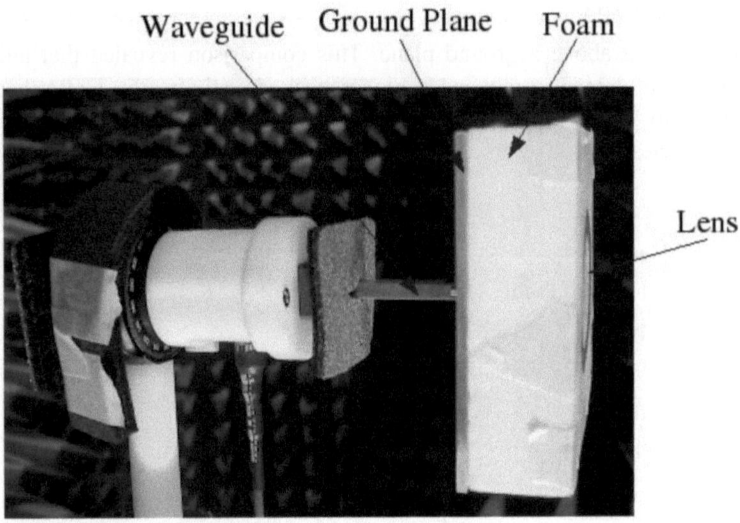

Fig. 3.12 Single-zone FZPA test set-up

3.3 Small 2 × 2 Single-Zone HFZP Array

As it was mentioned, by using arrays of smaller lenses instead of a single lens of equivalent aperture, the overall profile can be significantly reduced, making for a more compact design. Power amplifiers can be placed at the focus of each lens for power-combining applications.

When the array is used as an antenna, high grating lobe levels may result, since the spacing between lens elements is typically larger than a wavelength. Of course, hexagonal FZP printed lenses do not offer as high a radiation efficiency as their dielectric phase-correcting counterparts [1, 2]; however, they are much thinner, lighter in weight, easier to fabricate, less expensive, and can be used as a simple prototype to phase-correcting lenses. Moreover, a hexagonal shape is adopted instead of the conventional circular shape to improve the packing efficiency of the array [19, 20]. Figure 3.13 shows the geometry of single-zone hexagonal FZPL array.

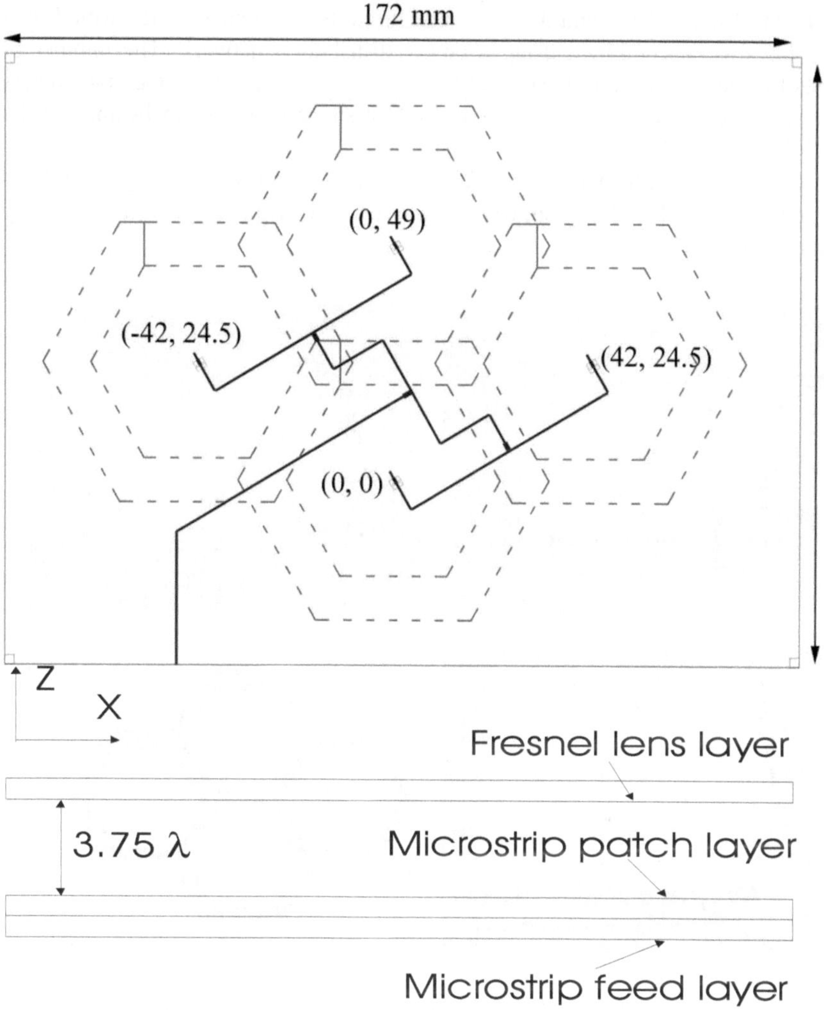

Fig. 3.13 Geometry of single-zone hexagonal FZPL array (after [20], courtesy of the Communications Research Centre, Canada)

The design of the hexagonal Fresnel plate is based on the circular zones of the conventional Fresnel lens. The circular zone of radius r_i was determined using classical formula (Eq. 1.17). The hexagonal zone most closely approximates the circular zone when it intersects it at 12 points, as it was shown in the Section 2.3. The half-height L_i of the ith hexagon can be related to the corresponding circular zone radius by:

$$L_i = 0.966 r_i .$$

Each of the FZPLs would require a separate feed antenna, which would then each have to be combined using a power distribution network. The number of zones for the FZPLs and the spacing between elements are the design parameters that are used to determine the directivity, the shape of the main beam, and the sidelobe / grating lobe levels.

As a proof of concept, a small 2 × 2 array of single-zone hexagonal FZPLs was designed at 30 GHz, as shown in Fig. 3.13 [20]. A prototype of a 2 × 2 array is shown in Fig. 3.14.

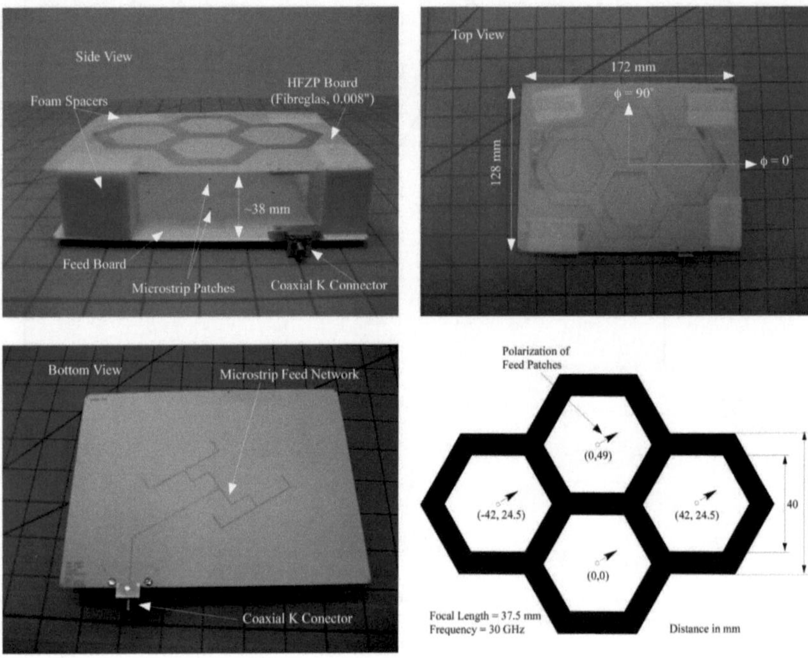

Fig. 3.14 A prototype of a 2 × 2 HFZP array (after [20], Courtesy of the Communications Research Centre, Canada)

The zones of adjacent lenses are made to overlap in order to reduce the inter-element spacing, which decreases the level of the grating lobes. The feed array consisted of four aperture-coupled microstrip patches printed on a 0.020-inch thick substrate with ε_r=3.5, while the microstrip feed lines were printed on a 0.010-inch-thick substrate (also with ε_r=3.5). The two substrates were then bonded together to form a single board. The FZPLs are printed on a thin layer of fiber glass substrate (0.008-inch thick) and are suspended above the feed network with foam spacers. The polarization of the patches was chosen to be at $\varphi = 30°$ (as shown in Fig. 3.14) in order to simplify the microstrip feed network layout.

3.3.1 Microstrip Feed Networks

Microstrip lines constitute the feed network used to excite each element in the array. Feed network design is dictated by the required relative phases and amplitudes of the electromagnetic waves radiated by each element in the array. If we want the array to have the maximum possible gain, the phases of the radiated electromagnetic waves from all the elements need to be uniform so as to produce the maximum radiation in a broadside direction.

A commonly used microstrip feed network is the parallel or corporate feed network. Such a feed network is characterized by successive divisions of a microstrip line to connect to the patch elements in parallel. A corporate microstrip feed network, like any feed network, will result in a certain amount of loss of antenna efficiency and gain, due to losses in the microstrip lines, undesired radiation from the lines, and mutual coupling between patches via surface waves.

According to the present design, a standard design principle to design the feed network was used [21]. The feed consists of a microstrip branch line fed in the center by a slot-coupled microstrip line located on a second substrate. A multi-layer approach was adopted to allow for the integration of active devices in a large array of parallel branches. By using the second layer, more area is made available for mounting the devices, and good isolation is provided to prevent any spurious radiation of the devices from interfering with the antenna pattern [22]. A measured pattern of the microstrip patch array at 28.6 GHz is shown in Fig. 3.15.

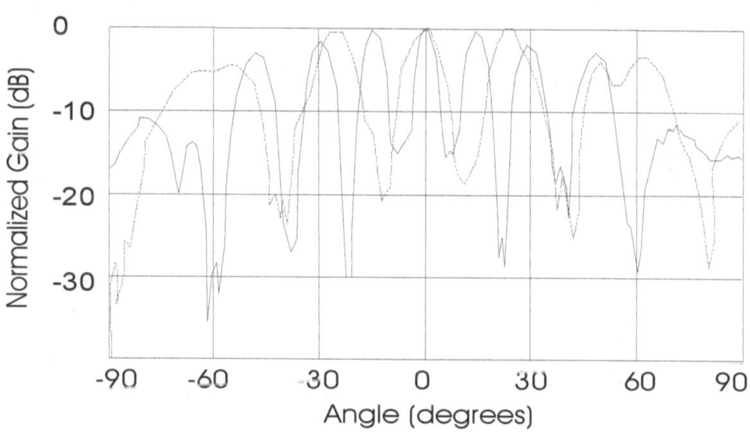

Fig. 3.15 Measured patterns of the microstrip patch array at 28.6 GHz: $\varphi = 0°$ (*solid* line) and $\varphi = 90°$ (*dashed line*)

3.3.2 Array Characteristics

Figure 3.16 shows the measured gain versus frequency of the prototype array and the return loss. The peak gain was expected to be at approximately 29 GHz (based on the FDTD simulation of a single hexagonal zone ring) while the measured peak occurred at 28.6 GHz. Beampattern of single-zone HFZP is shown in Fig. 3.17.

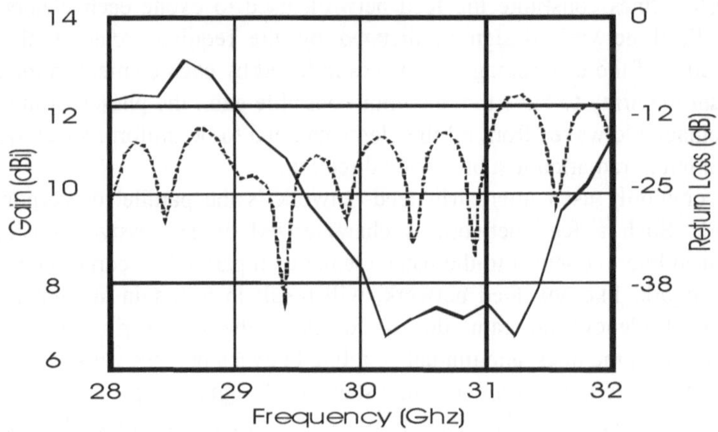

Fig. 3.16 Gain (*solid line*) and return loss (*dashed line*) versus frequency

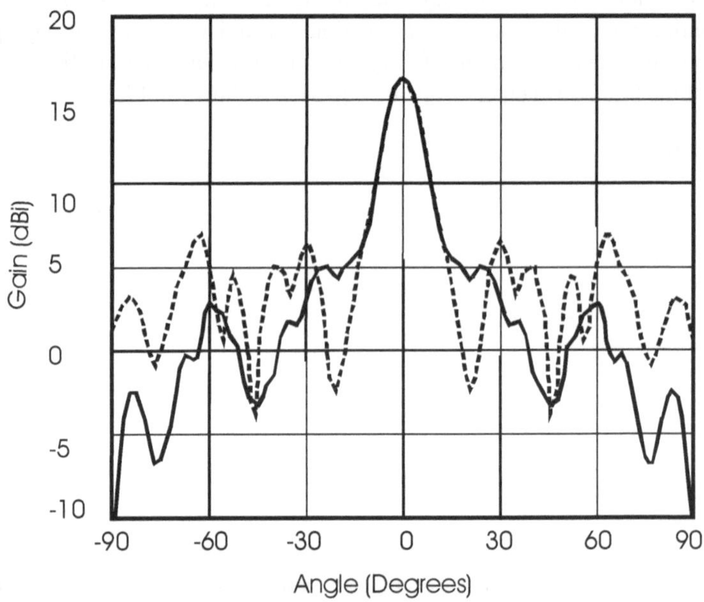

Fig. 3.17 Simulated pattern of the single HFZP (1 zone): $\varphi = 0°$ (*solid line*), $\varphi = 90°$ (*dashed line*)

There was a significant loss in the microstrip feed network because of the relative long transmission lines. Also there was a small amount of misalignment between the HFZP board and the microstrip board, which are the main reason for the antenna gain to be relative low.

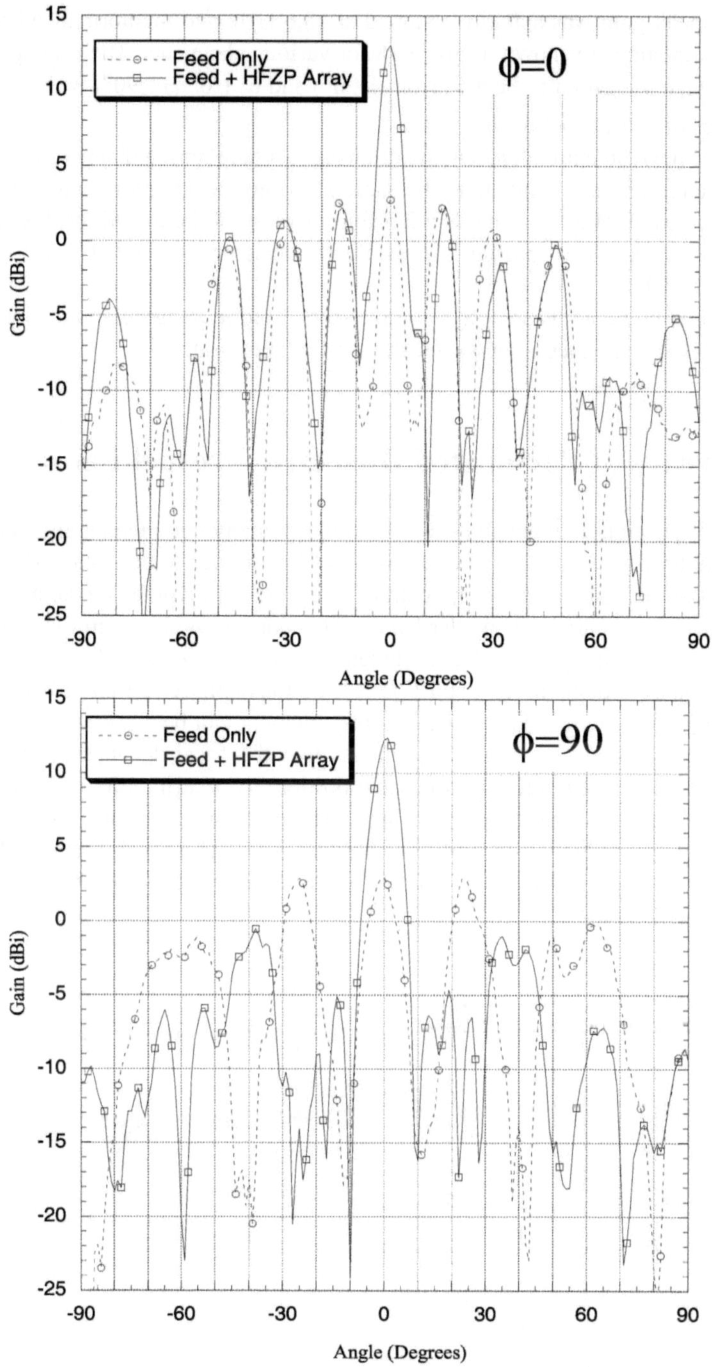

Fig. 3.18 Measured patterns at 28.6 GHz

The peak gain was actually obtained at 28.6 GHz and the normalized patterns at this frequency are shown in Fig. 3.18 for various plane cuts. The principal beam widths were measured to be 4.6° (for $\varphi = 0°$) and 6° (for $\varphi = 90°$) with maximum sidelobes of –11 dB below the peak.

To achieve an equivalent beamwidth using a single lens, a diameter of $D = 145$ mm is required, and maintaining the same F/D ratio of 0.64, the feed would have to be located at 94 mm, which is a 2.5-fold increase in the overall profile. Further reduction in the array profile is possible by designing the FZPLs with smaller values of F/D.

It could be noted these arrays could also be used in imaging applications since they will focus the incoming waves into a set of spot beams.

3.4 Arrays of Perforated Dielectric Fresnel Lenses

Low-profile, high-gain antennas are desirable in many commercial and consumer applications where esthetics is often of equal importance to antenna performance. One of the methods to reduce a depth of lens antenna involves replacing a single large lens with an array of smaller lenses. This concept is shown in Fig. 3.19 [23], which compares the profile of a single lens of diameter D and $F/D = 1$ to an array of four lenses of diameter $D/4$ and $F/D = 1$. So the depth reduction of lens antenna is made possible by the fact that each array element can be a portion, $d = D/x$, of the original antenna diameter. This means that the focal distance of each array element is therefore reduced by the same portion, $f = F/x$, in order to maintain the $F/D = f/d$ of the original lens. There are, of course, various tradeoffs that accompany the reduction in profile of the lens array. Increased losses in the feed network and degradation in the pattern due to potential grating lobes are among the important factors.

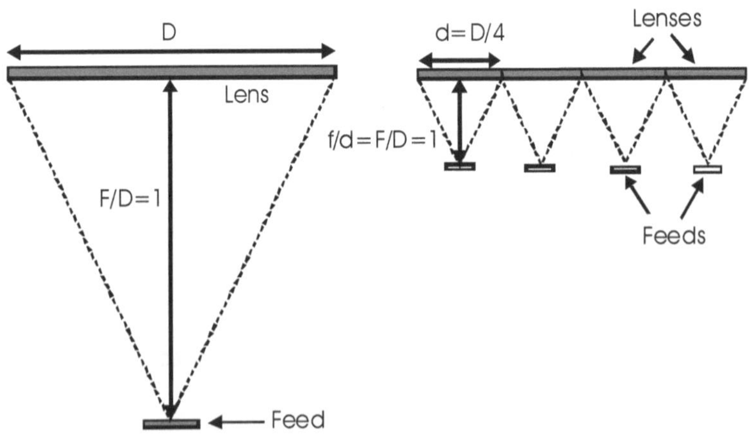

Fig. 3.19 Concept of depth reduction (after [23] with permission by Dr. A. Petosa)

3.4.1 2 × 2 Array of Perforated FZP Lenses

A 2 × 2 array of perforated dielectric Fresnel lenses was designed at 30 GHz with an $F/D = 0.25$ [23]. The geometry of the array is shown in Fig. 3.20. The overlap of the lenses was required in order to suppress the appearance of grating lobes. The lenses were fed with a 2 × 2 array of microstrip patch antennas where the feed layer was separated by a distance of 1.5 cm (1.5 λ) from the lens array. The overall profile of the lens array was 2.5 cm (approximately 1" or 2.5 λ).

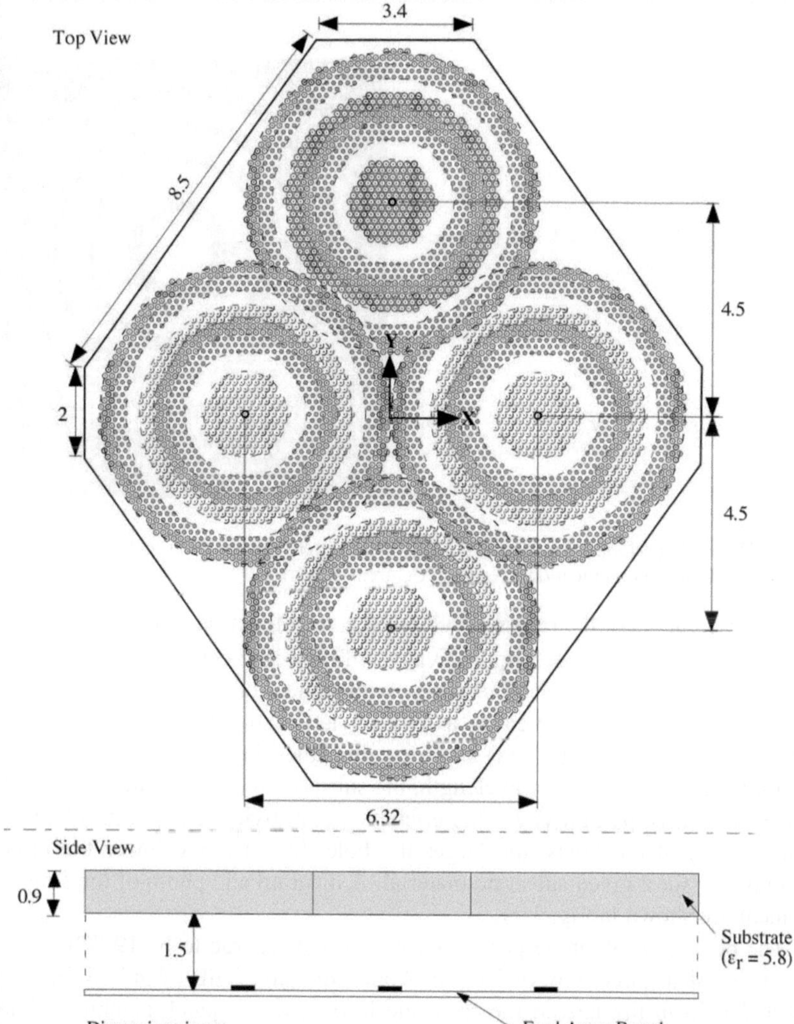

Fig. 3.20 Array of perforated dielectric Fresnel lenses designed at 30 GHz (after [23], courtesy of the Communications Research Centre, Canada)

The design of the single element was carried out at 30 GHz. An $F/D = 0.25$ was chosen, since the objective was to obtain a low-profile antenna [24]. The diameter of the lens element depends on the number of elements in the array and the overall array aperture.

The required values for the dielectric constants is related to the thickness of the lens [2, 25]:

$$t = \lambda_0 / \left\lfloor P\left(\sqrt{\varepsilon_i} - \sqrt{\varepsilon_{i-1}}\right)\right\rfloor,\ i = 2,3,\ldots,P.$$

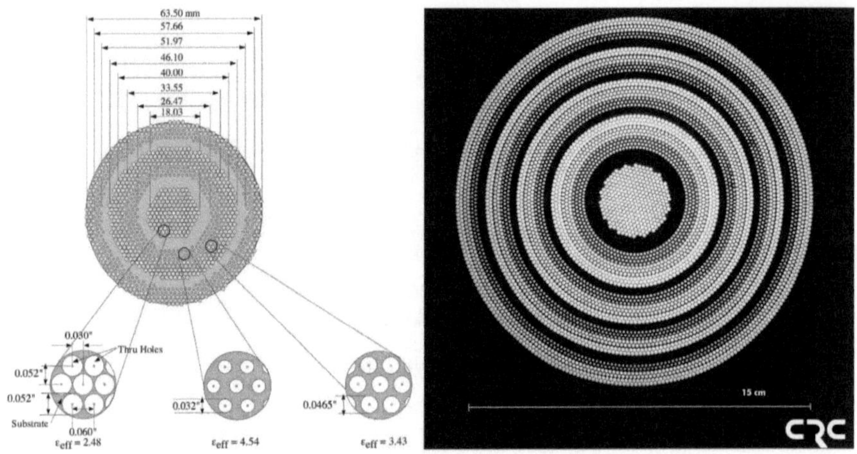

Fig. 3.21 Single-perforated dielectric Fresnel lens element design (*left*) and single lens (*right*) (Courtesy of the Communications Research Centre, Canada)

Generic curves of the required values of ε_i as a function of lens thickness were presented in [2, 25]. These required values of dielectric constant were obtained by perforating a single dielectric sheet with holes of varying diameters. The perforations are implemented as a uniform lattice of holes drilled through the substrate. If the separation (s) between the holes in this lattice is kept small compared to the operating wavelength, the substrate will appear to have a uniform effective relative permittivity. This effective permittivity can be controlled by the diameter (d) of the holes: the larger the hole diameter, the lower the effective permittivity for a given substrate material. A diagram and photo of the single lens element are shown in Fig. 3.21.

The peak gain of single-perforated lens was measured to be 19 dBi compared to a theoretical directivity of 23 dBi. Approximately, 2 dB of this difference can be attributed to the loss in the feed line leading to the patch antenna, while the other 2 dB can be attributed to the nonideal pattern of the patch antenna as well as mismatch loss due to reflections from the air–dielectric interface of the lens.

The normalized E- and H-plane patterns of the 2 × 2 lens array measured at 30.4 GHz are shown in Fig. 3.22, along with the predicted ideal patterns [24]. Despite the fact that the ideal patterns do not include the effects of the overlapping lenses, the agreement with the measured patterns is seen to be quite good.

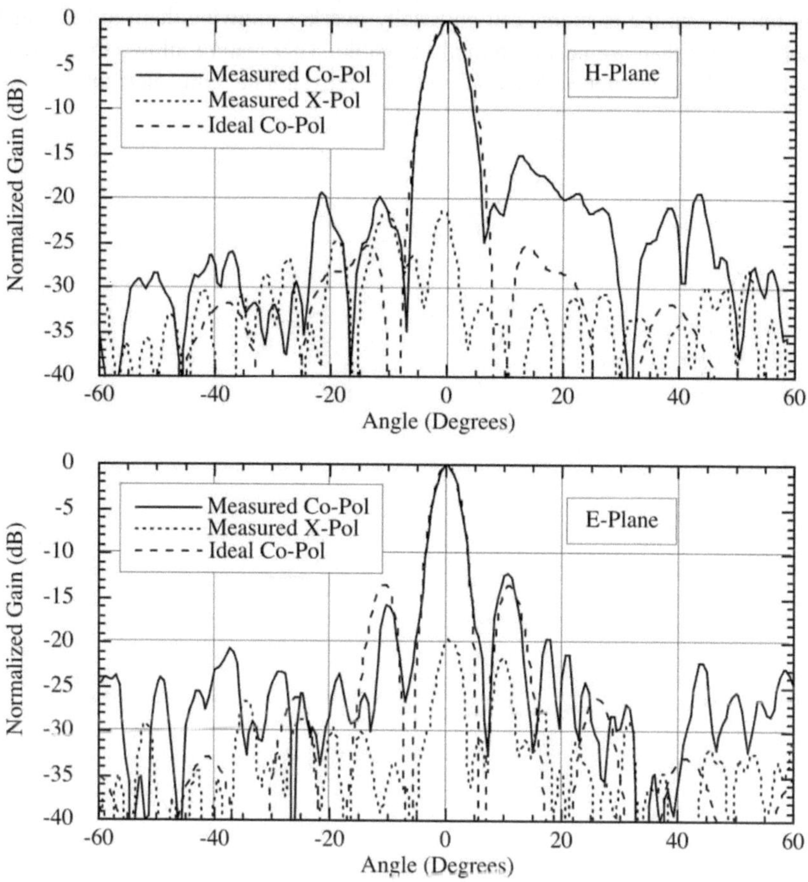

Fig. 3.22 Measured patterns of the perforated dielectric Fresnel lens 2 × 2 array at 30.4 GHz (after [23], Courtesy of the Communications Research Centre, Canada)

3.4.2 4 × 4 Array of Perforated FZP Lenses

Based on an array aperture of approximately 25 × 25 λ, the element diameter was chosen to be 6.35 λ, resulting in an array of 16 lenses, which was deemed sufficiently large to investigate the performance and to identify the challenges of this type of antenna configuration [24].

As with any array design, there are various ways of arranging the elements. Rectangular or triangular lattices are most commonly preferred for planar arrays. For the majority of planar arrays, low-gain antennas are used as the array elements, which are typically separated by a distance between $\lambda/2$ and λ, to avoid grating lobes. This array is somewhat uncommon in two respects: first, the elements have reasonably high gain and will be spaced several wavelengths apart, producing grating lobes in the array factor. The second aspect of this array design, which is quite rare, is that the elements can physically overlap, allowing them to be brought closer together than would be normally possible. This feature allows for a greater design flexibility, not normally associated with conventional array elements.

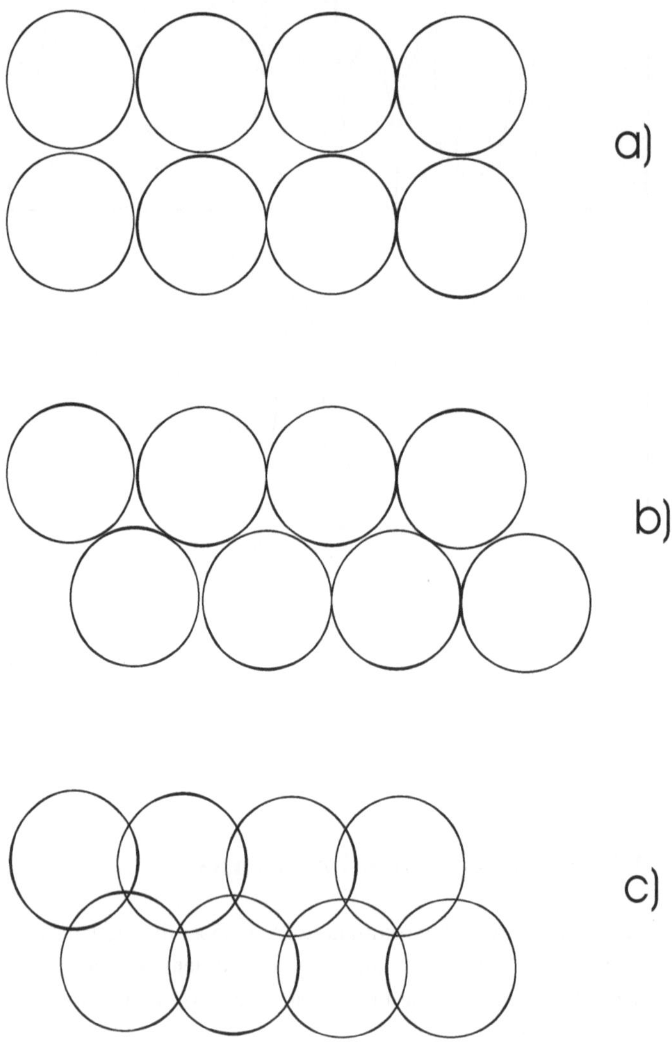

Fig. 3.23 Array lattice configuration (after [24] with permission by Dr. A.Petosa)

Three lattice configurations that were examined in [24] are shown in Fig. 3.23. Figure 3.23a and b show the elements arranged in traditional rectangular and triangular lattices. Figure 3.23c is also a triangular lattice, but the center-to-center spacing between elements is such that the elements overlap. The theoretical array pattern for these three cases were determined by using the measured pattern of the single-perforated dielectric Fresnel lens as the element pattern, and multiplying it with the array factor for each case. This approach ignored any coupling that might exist between elements and also ignored any phase errors that might arise due to the overlapping of the elements.

Patterns for the three lattices are overlaid in Fig. 3.24, along with the pattern of a single lens with a diameter of $D = 25\lambda$. Each array pattern is shown in the azimuth plane containing the worst-case grating lobe levels. The patterns are somewhat asymmetric due to slight asymmetries in the measured element pattern. The rectangular lattice results in the worst grating lobe levels (–7 dB below the peak), while the triangular array lattice with the overlapping elements achieves the best grating lobe levels (–12.5 dB). As expected, all the lens arrays have worse sidelobe levels than the single lens; however, by overlapping the lens elements, the grating lobes can be reduced to acceptable levels. The triangular lattice of overlapping elements was therefore selected for array design.

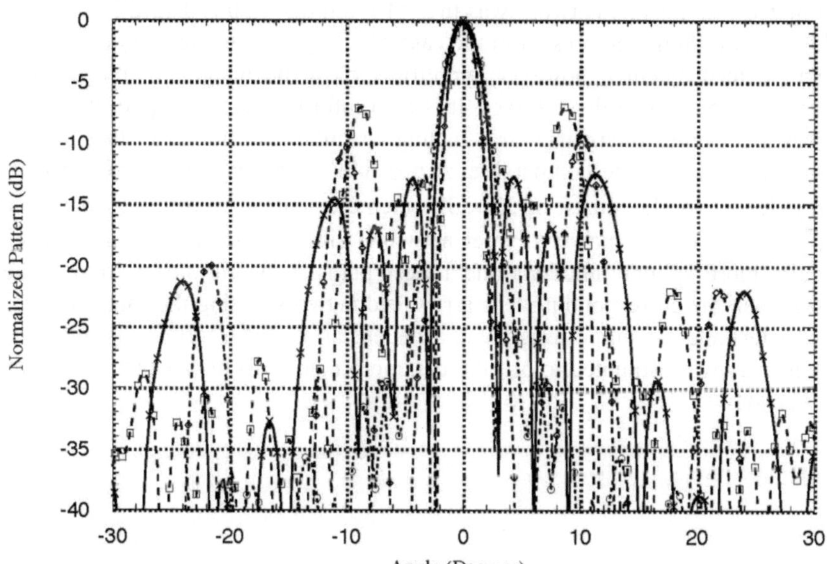

Fig. 3.24 Normalized theoretical array patterns

- - - - - (*circle*) - - - - - - Single lens ($D = 25\lambda$)
- - - - - - - (*square*) - - - - - ------Rectangular lattice array ($\varphi = 0°$)
- - - - - - - (*triangle*) - - - - - Triangular lattice array without overlap ($\varphi = 30°$)
________(X)________ Triangular lattice array with overlap ($\varphi = 90°$)

(after [24], with permission by Dr. A.Petosa).

In addition to the poorer sidelobe level performance of the array configuration, compared to the single lens, a second disadvantage is the increased losses due to the power distribution network required to excite the elements feeding the individual lenses.

The losses in the power distribution network can be divided into two components: one associated with the path length from the initial feed point to an element in the array and the second associated with spurious radiation from junctions. In general, the actual amount of power lost in the feed network is a function of the frequency of operation and the characteristics of the feed substrate. In this particular case, a simple T-junction power splitter was used, designed at 30 GHz on a 0.5 mm substrate having a dielectric constant of $\varepsilon_1 = 3.38$ and a loss tangent of 0.0027. The simulated insertion loss was found to be 0.6 dB. A 1-cm length of 50 Ω line printed on this substrate was simulated to have an insertion loss of 0.13 dB [26]. The insertion loss can be improved by optimizing the substrate thickness and permittivity and replacing the T-junction with a Y-junction, but this was beyond the scope of this investigation.

For the prototype described in the next section, the path length from the input of the power divider network to any of the lens feed elements is approximately 30 cm and includes four T-junction power splitters. The estimated loss for this network was thus (4×0.6 dB + 30×0.13 dB = 6.3 dB). This loss would not be present in the case of a large Fresnel lens illuminated by a single feed antenna; however, amplifiers could be integrated into the feed network to overcome these losses. It is also informative to compare these feed losses to a printed microstrip array having a similar aperture size, since the lens array profile would be comparable. A possible corresponding microstrip patch array design would consist of 32×32 patch elements spaced 0.75 λ apart and fed with a corporate power divider network. Since the microstrip array would require more splitter junctions and longer paths, it would suffer from higher losses than the corresponding lens array feed network. Using the same substrate parameters and T-splitter design as for the lens array, the insertion loss of such a patch array was estimated to be 10.6 dB, more than 4 dB higher than that of the lens array.

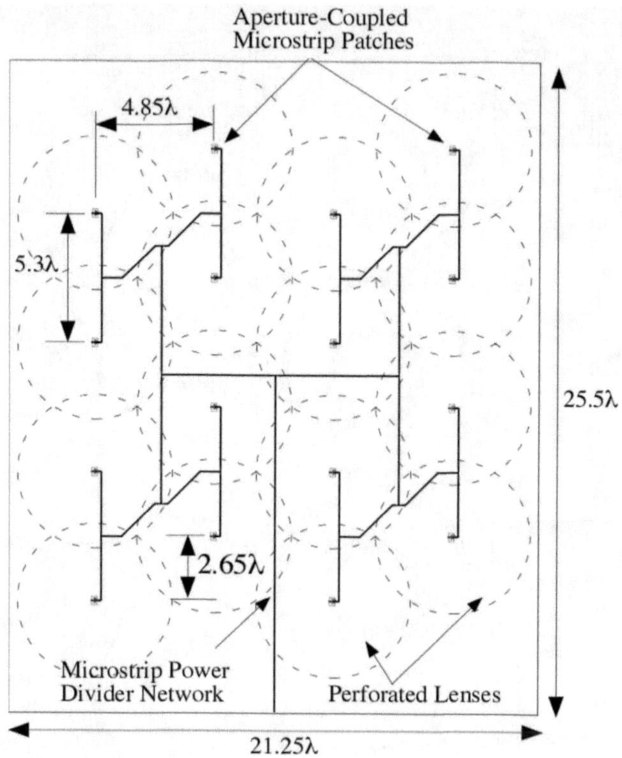

Fig. 3.25 Microstrip feed network for the lens array (after [24], Courtesy of the Communications Research Centre, Canada)

The array elements were arranged in a 4 × 4 triangular lattice, with a center-to-center spacing chosen so that the elements overlap enough to eliminate any gaps in the aperture. Figure 3.25 shows the power divider network for feeding the array of aperture-coupled patches and the location of each of the Fresnel lenses in the array. The patch array board was separated by a distance of 1.5 λ_0 from the lens array. The lens array was fabricated from a single sheet of material having a dielectric constant of 5.8 and a thickness of 0.9 λ_0. Thus, the overall profile of the antenna was approximately 2.5 λ_0 at 30 GHz, or 25 mm.

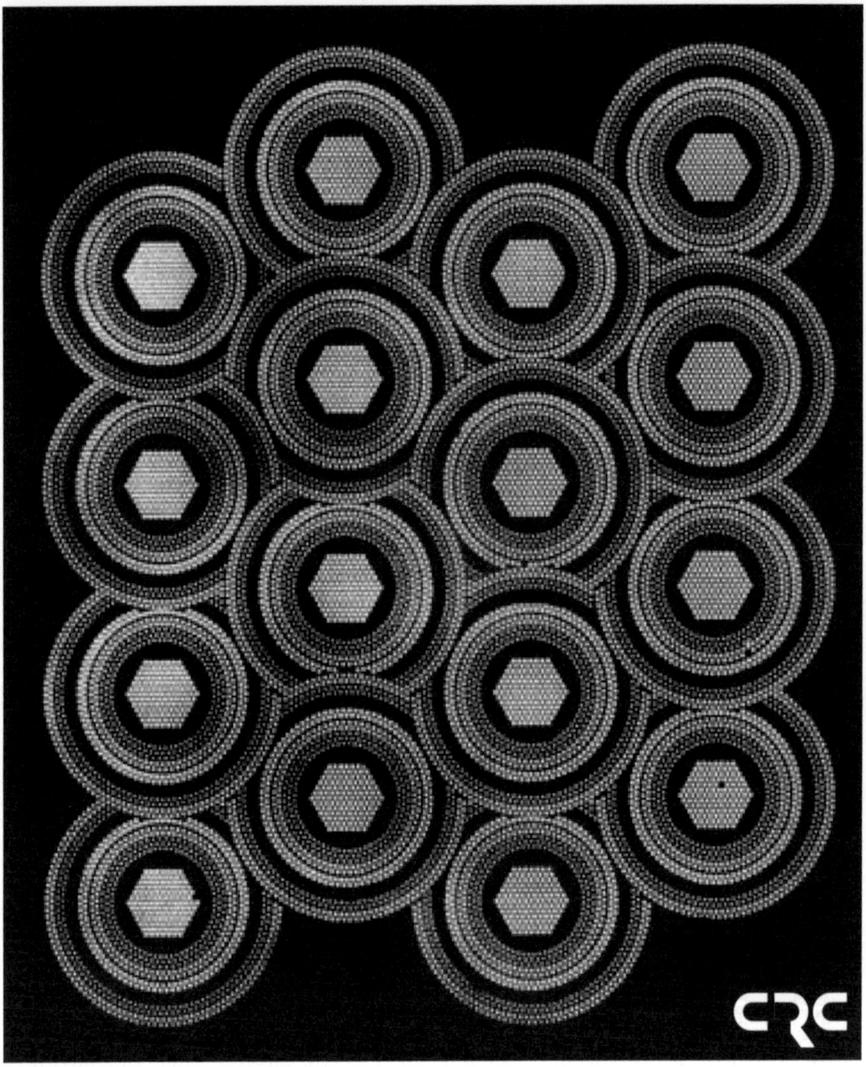

Fig. 3.26 Photograph of the perforated dielectric Fresnel lens array (after [24], courtesy of the Communications Research Centre, Canada)

A photograph of the lens array is shown in Fig. 3.26, and the assembled antenna along with the test fixture are shown in Fig. 3.27.

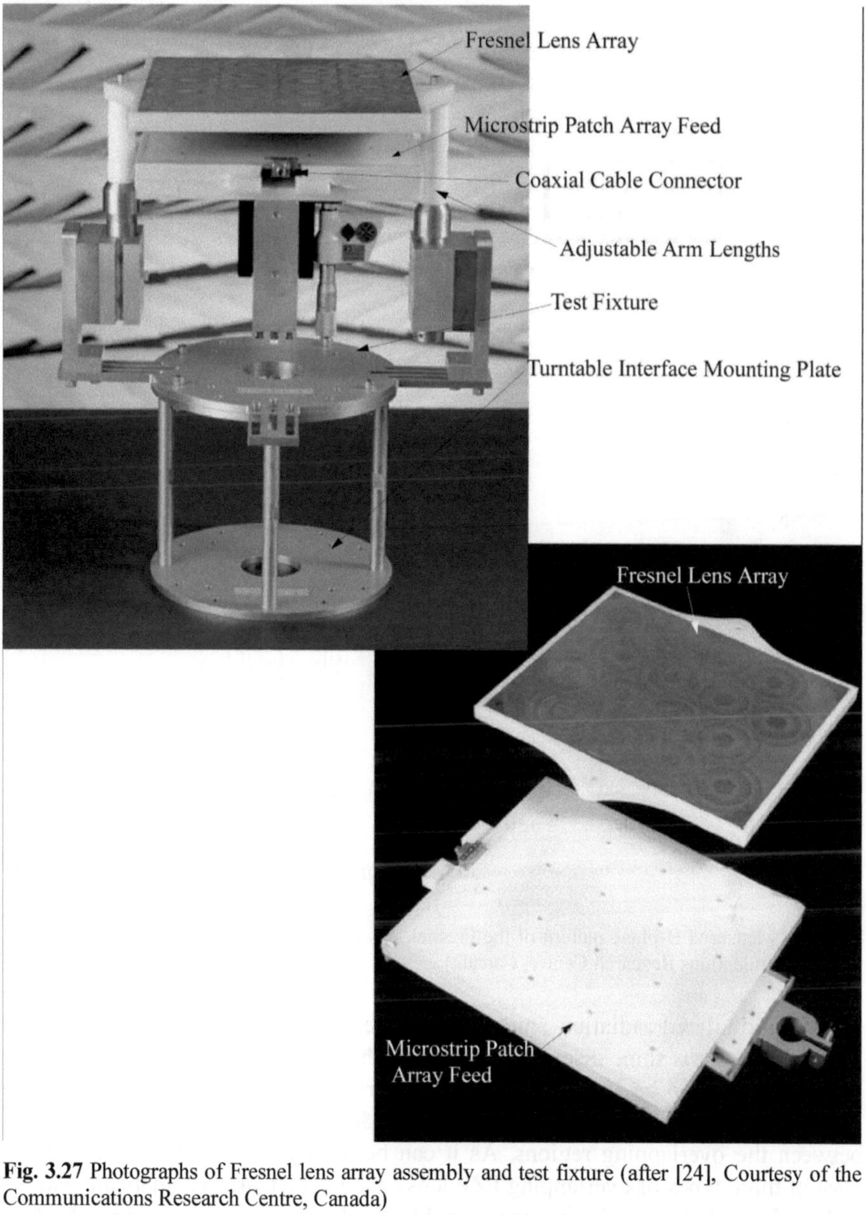

Fig. 3.27 Photographs of Fresnel lens array assembly and test fixture (after [24], Courtesy of the Communications Research Centre, Canada)

Figure 3.28 shows the measured H-plane co- and cross-polarized patterns of the Fresnel lens array at 28 GHz, where a peak gain of 18.5 dB was measured. The gain peak occurred at a lower frequency than the design frequency of 30 GHz and may be attributed to the poor performance of the feed network as mentioned above.

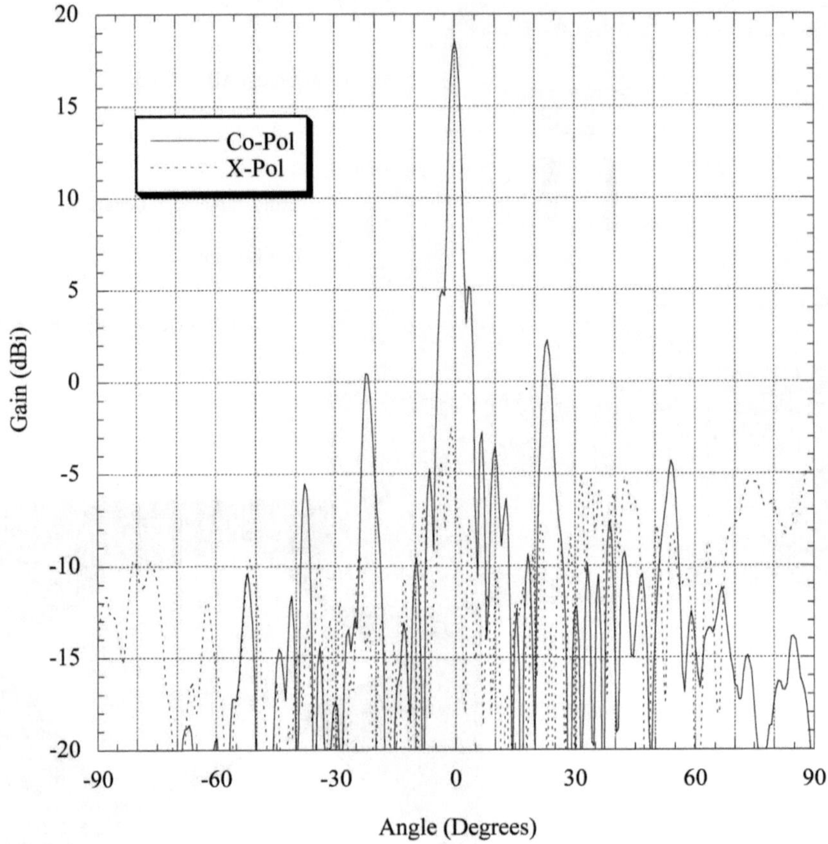

Fig. 3.28 Measured H-plane pattern of the Fresnel lens array at 28 GHz (after [24], Courtesy of the Communications Research Centre, Canada)

The normalized radiation patterns of the lens array at 30 GHz are shown in Fig. 3.29. There is some discrepancy between the ideal and the measured patterns.

There may also be some errors that arise from the overlapping regions of the individual Fresnel lenses. The ideal patterns do not account for any interaction between the overlapping regions. As it can be seen from Fig. 3.26, there are at least of three types of overlapping FZP lens shape, which are schematically shown in Fig. 3.30. Any such interactions would introduce phase errors across the lens aperture, which would then result in the asymmetries in the sidelobe levels which are seen in the measured patterns.

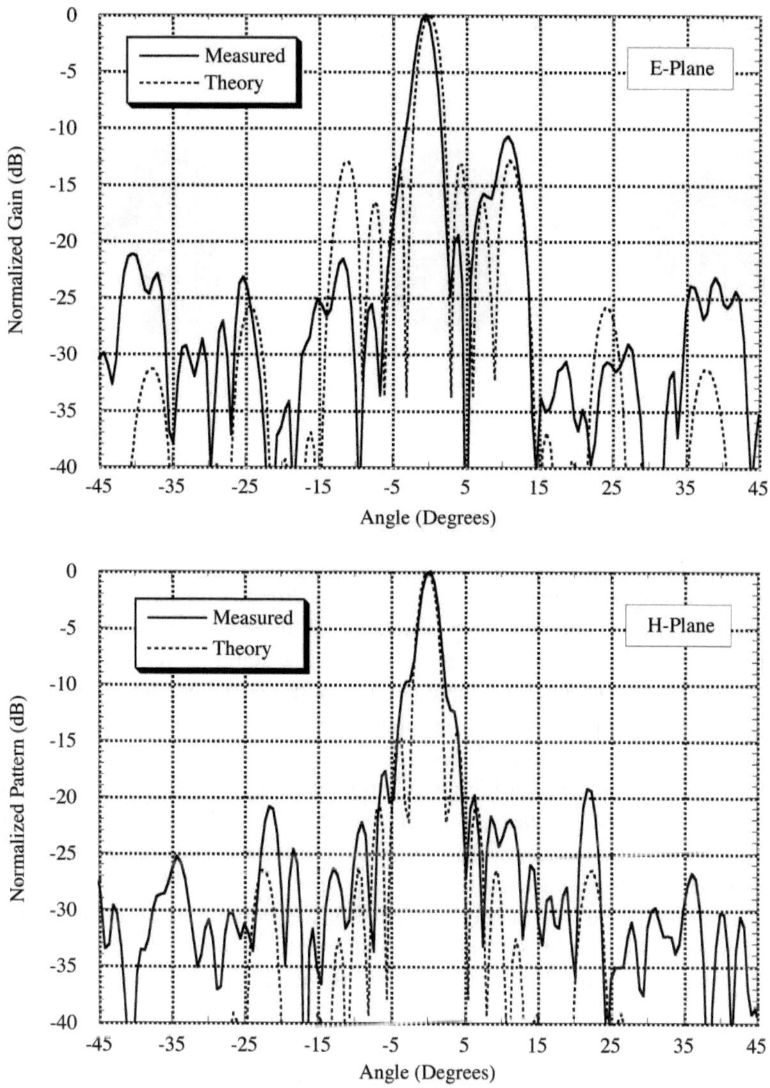

Fig. 3.29 Normalized patterns of the perforated dielectric Fresnel lens array at 30 GHz (after [24], courtesy of the Communications Research Centre Canada)

Fig. 3.30 Three main types of overlapping FZP lens shape

Thus by using perforated dielectric Fresnel elements, greater control over the grating lobes can be achieved due to the fact that the lens elements can physically overlap, thereby decreasing the center-to-center spacing to reduce grating lobe levels. The amount of overlap achievable is a tradeoff between a reduction in grating lobe levels and an increase in phase errors due to the mutual interference of the individual lenses, and a balance must be achieved.

Although it will not have as high a radiation efficiency as a corresponding single Fresnel lens due to the increased losses in the feed network, amplifiers can always be integrated into the feed network to offset these losses. The concept of using an array of Fresnel lenses to lower the antenna profile is found to be feasible, and this type of array would be well suited for spatial power combining applications where a much higher EIRP could be obtained compared to a single lens design by using several high-power amplifiers integrated into the feed of each lens element [25].

3.5 Simple Circular Zone Determination in Overlapping Case

3.5.1 Linear Two-Elements FZP Array

The circular zone radii for each FZPA were determined using Eq. (1.17) with a focal distance of 0.75λ at 30 GHz. The zones for each FZPA were placed, as computed, in the aperture plane (i.e. 0.75λ from the open waveguide) with a separation of 3.162λ.

In order to study the effect of the center zone being blocked, the three metal zone case was modeled using two different methods in FDTD. The first method used a ring subtraction technique, which meant that the zones were created by subtracting the air zones from the metal zones, leaving only the metal zones floating in air. With this approach, wherever there is metal, metal will be simulated. The second method used a priority technique, which meant that each zone was assigned a priority with higher priority zones cutting holes in the lower priority zones. In this approach, the outer zone had the lowest priority and every zone moving inward was assigned a progressively higher priority. In this way, the outer zones from one lens that would normally be blocking the center zone of the other lens would no longer be blocking since the center zone had a higher priority than the outer zones and would remain as air. Figure 3.31 shows a sketch of how these two approaches affected the resulting geometry in the three-zone case. Using the priority approach yields zones that more closely resemble the "modified" zones shown in Fig. 3.31(b). The modelings were prepared by Ph.D. student Ms. Sara M. Stout-Grandy.

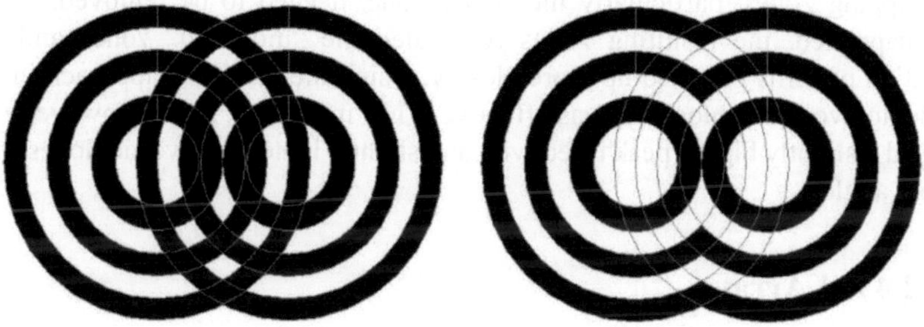

Fig. 3.31 Ring subtraction approach (*a*), priority approach (*b*)

Figure 3.32 shows the normalized radiation patterns for the three metal zone case. Figure 3.32 (a) shows the circular zone case with the ring subtraction approach whereas Fig. 3.32 (b) shows the circular zone case with the priority approach. The peak directivity of the "modified" zone case was about 0.8 dB higher than the circular zone case when using the ring subtraction approach where the center zone was partially blocked. However, when the center zone was not blocked in the priorities approach case, the directivity difference dropped to about 0.1 dB. The overall sidelobe level of the "modified" zone case was about 1.1 dB better than the circular case using ring subtraction, but this difference dropped to about 0.6 dB for the priorities approach.

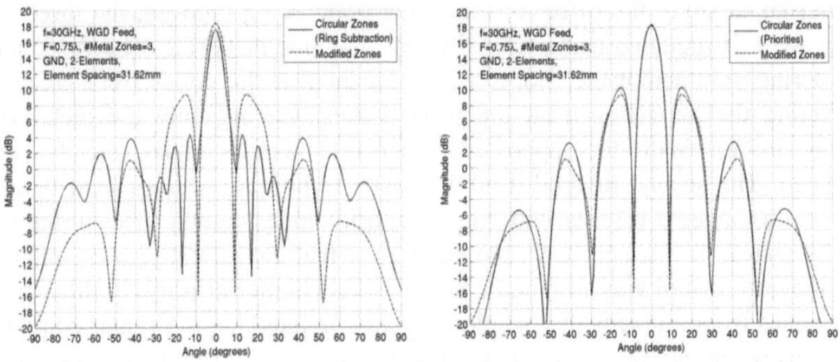

Fig. 3.32 Three metal zones with ring subtraction approach (*a*) and with priorities approach (*b*) for circular case (with permission by Dr. Sara M. Stout-Grandy)

The three-zone circular elements, without modification to their overlap areas, do not yield comparable radiation patterns and directivity. In order to make use of the simpler-to-model circular zones, open areas that were being blocked by overlapping zones, particularly the center zone, needed to be removed. When this happened, the resulting zones resembled the "modified" zones and the radiation patterns matched up more closely. Using the priorities approach in the model to avoid this blockage was effective. The "modified" zones, however, still yielded a slightly higher peak directivity and slightly better relative sidelobes over all phi angles.

3.5.2 4 × 4 Array

Prior to modeling the 4 × 4 array, a grating lobe analysis was again undertaken to determine an appropriate element spacing to maintain the grating lobes better than 10 dB below the peak. The 4 × 4 array geometry was set as an equilateral triangle lattice and the analysis was performed for two different focal distances, $F = 0.75\,\lambda$ and $F = 1.25\,\lambda$, both with two metal zones. The analysis showed that, in the absence of mutual coupling and interference effects, the element spacing at 30GHz should be about $2\,\lambda$ for the $F = 0.75\,\lambda$ case and $2.5\,\lambda$ for the $F = 1.25\,\lambda$ case.

Table 3.5 summarizes the results for the 4 × 4 simulated FZPA arrays.

Table 3.5 Summary of simulation results

Focal distance (λ)	Zone type	# Metal zones	Element spacing (λ)	Peak directivity (dB)	Max. relative sidelobe level (dB)
0.75	Circular	2	2	22.68	–4.96
1.25	Circular	2	2.5	24.56	–5.41
1.25	Circular	2	3	25.87	–10.16

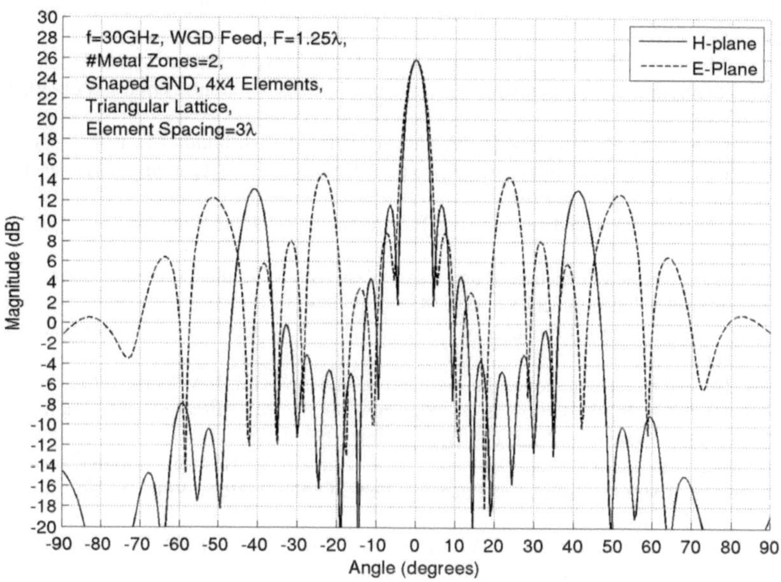

Fig. 3.33 The radiation patterns of the 4 × 4 array with $F = 1.25\lambda$ and two metal zones with 3λ element spacing (with permission by Dr. Sara M. Stout-Grandy)

Figure 3.33 shows the radiation patterns of the 4 × 4 array with $F = 1.25\,\lambda$ and two metal zones with 3 λ element spacing.

Thus using equilateral triangular array lattices, it was found that a low-profile 4 × 4 array could yield nearly 25 dB directivity and overall grating lobes better than 10 dB down from the array peak.

References

1. A. Petosa, A. Ittipoon, S. Thirakoune, Array of Perforated Dielectric Fresnel Lenses. URSI International Symposium on EM Theory, EMTS 2004, Pisa, Italy, May 23–27, 2004, pp. 969–971.
2. O. V. Minin, I. V. Minin. Diffractive Optics of Millimetre Waves. IOP Publisher, Boston-London, 2004.
3. P. F. Goldsmith, Quasioptical Systems, IEEE Press, New York, 1998.
4. R. Guenther, Modern Optics, John Wiley & Sons, New York, 1990.
5. P. D. Kearey, and A. G. Klein, Resolving Power of Zone Plates. Journal of Modern Physics, Vol. 36, No. 3, 1989, pp. 361–367.
6. R. C. Johnson (Ed.). Antenna Engineering Handbook, 3rd ed., McGraw-Hill, Inc. New York, 1993.
7. I. V. Minin, O. V. Minin, N. Gagnon, A. Petosa. Investigation of the Resolution of Phase Correcting Fresnel Lenses with Small Focal Length –to-Diameter Ratio and Subwavelength Focus. Proc. Electromagnetic Theory Symposium (EMTS) 2007, Canada July 26–28, Ottawa (URSI).
8. I. V. Minin, O. V. Minin, N. Gagnon, A. Petosa. Investigation of the Resolution of phase correcting FZP lenses with Subwavelength focus. Computer Optics, Vol. 30, 2006, pp. 65–68.
9. Y. J. Guo, S. K. Barton, T.M.B. Wright. Focal Field Distribution of Fresnel Zone Plate Antennas. Proc. International Conference on Antennas and Propagation – ICAP 1991, York UK, April 1991, pp. 6–8, 1991.
10. C. A. Balanis. Advanced Engineering Electromagnetics. John Wiley & Sons, New York, 1989.
11. G. B. Airy. On the Diffraction of an Object Glass with a Circular Aperture. Transactions of the Cambridge Philosophical Society, 1834, 5, p. 283.
12. M. Born and E. Wolf, Principles of Optics, 5th ed., Pergamon Press, Elmsford, NY, 1975, (Subsect 8.6.2).
13. J. W. Goodman, Introduction to Fourier Optics, McGraw-Hill, New York, 1968.
14. G. I. Greisukh, S. T. Bobrov, S. A. Stepanov. Optics of Diffractive and Gradient-Index Elements and Systems. SPIE Press, Bellingham, 1997, 414p.
15. R. Oron, J. L. Guedalia, N. Davidson, A. A. Friesem, E. Hasman. Anomaly in a High-Numerical-Aperture Diffractive Focusing Lens. Optics Letters, Vol. 25, No. 7, April 1 2000, pp. 439–441.
16. S. M. StoutGrandy, A. Petosa, I. V. Minin, O. V. Minin, J. S. Wight, Investigation of Low Profile Fresnel Zone Plate Antennas. Accepted for publication in Microwave and Optical Technology Letters 2008. See also: S. M. StoutGrandy, A. Petosa, I. V. Minin, O. V. Minin, J. S. Wight. Investigation of Low Profile Fresnel Zone Plate Antennas. URSI North American Radio Science Meeting, Ottawa, Ontario, July 23, 2007.
17. R. C. Johnson. Antenna Engineering Handbook, 3rd ed. McGraw-Hill, New York, 1993.
18. G. Z. Aizenberg, V. G. Yampolski, O. N. Tereshin. Microwave Antennas, Part two, Moskow: Svyaz, 1977 (in Russian).
19. S. Stout-Grandy, A. Petosa, I. V. Minin, O. V. Minin, J. S. Wight. Hexagonal Fresnel zone Plate Antenna'. Proc. International Symp. On Antenna Technology and Applied Electromagnetics (ANTEM 2006), Montreal, Canada, July 2006.
20. A. Petosa, S. Thirakoune, I. V. Minin, O. V. Minin. Array of Hexagonal Fresnel Zone Plate Lens Antennas. Electronics Letters, Vol. 42, No. 15, 20th July 2006, pp. 834–836.
21. R. Garg, P. Bhartia, I. Bahl, I. B. A. Ittiboon. Microstrip Antenna Design Handbook, Artech House, London, 2001. (Chapt. 12).
22. A. Petosa, R. Larose, A. Ittipiboon, M. Cuhaci. Microstrip-fed Array of Multisegment dielectric Resonator Antennas. IEE Proc.-Microw. Antennas Propag., Vol. 144, No. 6, December 1997, pp. 472–472.

23. A. Petosa, A. Ittipiboon, S. Thirakoune. Array of Perforated dielectric Fresnel lenses. URSI Int. Symp. on Electromagnetic Theory (EMTS 2004), Pisa, Italy, May 2004, pp. 969–971.
24. A. Petosa, S. Thirakoune, A. Ittipiboon. Investigation on arrays of perforated dielectric Fresnel lenses. IEE Proc., Microw. Antennas Propag. Vol. 15, No. 1, 2006, pp. 7–9.
25. A. Petosa, A. Ittipiboon, S. Thirakoune. Design and Performance of a Perforated Dielectric Fresnel Lens", IEE Proceedings Microwave Antennas Propag., Vol. 150, No. 5 Oct. 2003, pp. 309–314.
26. Ansoft Ensemble Version 8, www.ansoft.com

Chapter 4
Some Fields of Lens Array Applications

4.1 Array of Small Lenses

Usually, the grating lobe criteria [1, 2] that restricts the scan angle of a phased array to eliminate grating lobes from entering "real space," as it was mentioned in Sect. 1.2.10, is defined as

$$\sin\theta \le (\lambda / d) - 1,$$

where θ is the scan angle of the array, λ the operating wavelength, and d the spacing between elements. Such a condition eliminates any beam, other than the main beam, from forming in visible space, and the antenna designer need not be concerned about radiating energy or receiving energy (other than normal sidelobes) from unwanted directions. This is a desirable condition (no grating lobes in real space) but imposes a strict limit on element spacing. For a nonscanned antenna – main beam on boresight – the theoretical element spacing can approach λ. For large scan – approaching 90° – the element spacing approaches a $\lambda/2$ limit. So the condition on the spacing between elements is:

$$d \le \frac{\lambda}{\sin\theta + 1}.$$

In a phased array with limited-scan capability, the radiating element spacing is larger than one half wavelength and the element gain is higher than its full-scan counterpart to avoid grating lobes. The array element proposed in [3] was low-loss and can provide any gain required since the size of the lens can be as large as needed.

There are several methods used to increase the gain of a small array element. The most common method is to sub-array several elements together. Although this is a very good approach in the microwave range, it becomes very lossy at millimetric wavelengths because of the feed network connecting the elements of the sub-array. Another technique is to place the radiating element in a metal cavity [4]; however, the system becomes heavy and bulky. Dielectric lenses have also been used in conjunction with radiating elements, e.g. extended hemispherical lenses directly located on a slot antenna [5] or a dielectric lens directly located on a microstrip patch [6]. In the latter case, those lenses offer limited gain increase because of their small size.

Gains were measured for the plano-convex lens in various configurations [3]. The single patch gain was about 5 dB, but with the lens it increased by 8 dB to about 13 dB. The four-element array gain without the lenses was about 8.5 dB and increased by 8.5 dB to about 17.5 dB with the lenses. Using the meniscus lens instead, the gain increase was about 1 dB more than for the plano-convex lens. The main reason is that the meniscus lens has less spillover because of its shape. The lenses were not surfaced matched and the gain change caused by this mismatch was observed as much as $\pm$1.1 dB, which shows that surface matching would be beneficial.

The excitation of dielectric spheres by waveguides [7] and the probe excitation of a dielectric hemisphere [8] have been investigated. The near fields of the patch can be a significant factor in the excitation process and were modeled approximately as dipole line sources at the patch extremities in [9]. Experiments on a range of different size spheres indicated that increasing the sphere diameter beyond about $D = 3\lambda_0$ (λ_0 = free-space wavelength) yields little increase in gain. This is attributed to the inability of the patch to create an optimal field distribution. Shaping of the spherical surface and using different shapes of patch resonators are some of the many variations that offer additional pattern control. In general, the presence of the sphere brings about improved equality of E and H beamshapes, a reduction of sidelobes, and some reduction in cross-polarization levels [9].

The use of spheres to effect pattern control is most marked when considering the arrays of these elements [9]. Without the spheres, the patch spacing d was made too great for conventional array operation and hence the pattern was dominated by grating lobes. With the spheres in place, a well-structured beam with sidelobes commensurating with a uniform array distribution was achieved, which is in very good agreement with theory. Cross-polarization levels were lower than those usually obtained from microstrip arrays, where feed radiation is a problem. Further details on such array and another version with closer patch spacing are summarized in Table 4.1, which further illustrates the properties of the new element.

Table 4.1 Antenna array parameters measured at 90 GHz (after [9])

	256 Element		64 Element	
	Without spheres	With spheres	Without spheres	With spheres
$d/\lambda_0 = D/\lambda_0$	0.95	0.95	1.90	1.90
Gain, dBi	24	27	14	26
Spurious (H), dBi	<–20	<–17	<–17	<–20
Spurious (E), dBi	<–11.5	<–17	<–18	<–23
Cross-polar (H), dBi	<–15	<–16	<–12.5	<–20
Cross-polar (E), dBi	<–25	<–23	<–15	<–24

One of the problems with arrays of high-gain elements is that they have a very limited beam scan range capability, due to the appearance of grating lobes because of the wide element spacing. These arrays are therefore usually only used for fixed beam applications [10].

4.2 Lens Array Based on Luneberg Lenses

A single Luneburg lens on an azimuth mount has few advantages, as it is easier to move the feed. However, a multiple lens design provides significant advantages in overcoming shadowing problems. Because no significant azimuth scanning is needed, the lenses can be closely packed horizontally. By stacking these rows on a tilted plane, close packing in an orthogonal direction is possible. The whole tilted plane is now effectively a filled aperture with a low-frequency cutoff defined by the dimensions of the array. Thus, very small lens can be used saving total weight and without compromising the low-frequency performance. Use of small lenses also allows operation at higher frequencies due to reduced dielectric loss. This concept is also applicable to parabolic reflector designs, where it is known that reflector cost per unit area decreases with reflector size. Luneburg lens is such a design that allows the synthesis of large apertures at low cost.

As a general guide, the authors [11] were considering the full Luneburg lens with diameters of around 5 m or less and for the virtual source lens diameters are generally 8 m or more. They believed that the possibility of having several simultaneous beams on the sky would give the SKA (Square Kilometre Array Radio Telescope) a unique and very attractive feature for a new radiotelescope for the 21st century. In a station composed of 176 elements of 7-m diameter, the corresponding station diameters are 500 and 250 m for elevation limits of 15° and 30°, respectively (Fig. 4.1). With this composition, each station is equivalent in collecting area to a 93-m diameter dish.

Fig. 4.1 Concept of SKA images with array of Luneberg lenses (Photo credit: "University of Sydney Vislab" with permission by Helen Sim, and Images courtesy Chris Fluke, Centre for Astrophysics & Supercomputing, Swinburne University of Technology)

4.3 Array of Luneberg Lenses Protrudes Outside of the Fuselage

The invention [12] utilizes dielectric lenses in the form of an array of hemispherical lens mounted on a ground plane to focus into a pencil or fan beam the energy radiated from an array of point sources located near the surfaces of the hemispheres. When mounted upon the fuselage of an aircraft, the lens array has an advantage over a single sphere in free space, in that each hemispherical lens extends only one half as far outside of the fuselage and into the air stream as compared to a complete spherical lens. Furthermore, because the antenna consists of an array of hemispheres instead of a single hemisphere having the same gain as the array of hemispheres, the array of hemispheres protrudes outside of the fuselage a lesser amount than would the single hemisphere having the same gain. For these reasons, the hemispherical lens array is a "low-profile" antenna.

Although the beam generated by the array of hemispherical lenses may be in the form of a "pencil" beam or a "fan"-shaped beam. It should be understood that actual shape of the beam generated by the array will depend on the relative dimensions of the hemispherical lenses, the number of lenses and point sources, the spacing of the lenses in the array, and the manner in which the lenses are illuminated by the array of point sources.

The prototype low-profile antenna [13] consists of four lens hemispheres mounted on a ground plane. The lenses are phase-combined to produce a beam. The beam is varied in elevation by the movement of the feed assembly and in azimuth by the movement of the ground plane.

The array parameters are as follows:

- Type of lenses: Luneberg 4-lens array
- Frequency: 19.2–21.2 GHz
- Polarization: LHCP or RHCP (selectable)
- Beamwidth (nominal): 2° (AZ) – 5° (EL)
- Diameter: 762 mm
- Height: 152 mm

4.4 Fly's-Eye Imaging Concept

Arraying of lenses is used extensively at optical frequencies for imaging applications [14, 15]. In these cases, the lens array elements are typically being used to create a series of spot beams where detectors are placed to pick up the light and create an image. Note that at microwave frequencies, antenna arrays described in the Chap. 3 are used to enhance gain by focusing the radiation of all elements to the same point. Instead of using only a single-lens antenna, it is also possible to put lens array in the focal plane of an objective lens. The resulting configuration is shown in Fig. 4.2, where the objective lens is either placed in the Fresnel zone or in the far field of the lens antenna (eye lens).

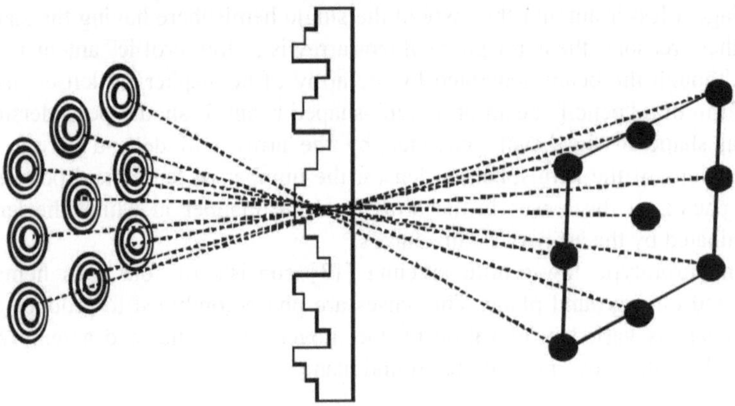

Fig. 4.2 Fly's-eye imaging concept

For imaging, it is important that a sharp image of a certain object is obtained and this means that an incoming plane wave has to be focused to the planar feed of one of the lens antennas. In the fly's-eye concept, the focusing is achieved in two steps. First, the objective lens transforms the incoming plane wave into a (nearly) spherical wave propagating toward an eye-lens antenna. Then, this spherical wave is focused to the phase center of the eye lens.

4.5 Waveguide Lens Array Technology

Waveguide lens array technology uses waveguide elements as a constrained refractive media. The phase velocity in waveguide exceeds the speed of light, and the lengths of waveguide between the lens surfaces are chosen to produce a collimated aperture distribution. The waveguide is a dispersive media, i.e. the phase velocity varies with frequency. As a consequence, the operating bandwidth of these designs is limited. Zoning techniques, i.e. truncating the waveguides to achieve the required aperture distribution in a modulo 2π sense, can actually enhance the bandwidth of the lens as mentioned in [16].

A so-called focusing pseudo-lens with variable gradient of refractive index was discussed in [17] (see Fig. 4.3). In this case, the artificial metal-dielectric was a waveguide structure – the lens consists of array of circular waveguides with different diameter vs. radius. The following relation gives the equivalent refractive index of the waveguide structure as a function of the radius r of the waveguide cross section:

$$n = c/v = \left[1-(\lambda/3.42R)^2\right]^{1/2}.$$

The refractive index of the waveguide structure in the converging pseudo-lens changes in the radial direction from 0.95 for R = 30 mm (radius of the central waveguide cross section) to 0.37 for R = 10 mm. At the wavelength λ = 3.2 cm, the thickness of the pseudo-lens was 5.3 cm.

Fig. 4.3 Microwave lens made from an array of circular waveguides (after [17])

Modification of such a waveguide lens array was used in so-called Cybertenna [18] – see Fig. 4.4. The following main modifications were used: the diameter of waveguides was fixed but the length of waveguides was varied; the circular waveguides were changed to rectangular. The Cybertenna was designed as a "zero thickness waveguide lens". The developed technology was called WAVEGUIDE LENS TECHNOLOGY™ [19], where a microwave lens antenna consists of an array of 680 rectangular waveguides, arranged rotationally symmetrically around the antenna axis. To reach the goal of a mass-producible antenna, waveguides are formed by a low epsilon material as the inner volume of the waveguide, and a layer of conductive paint placed on the outside of these formed waveguides. The material chosen is expanded polystyrene [18]. The antenna operates as a concave lens, with an index of refraction $n < 1$, which focuses the signals at a focal plane behind the antenna. The concave properties of the lens ensures that the received plane waves are focused at one point. The phase is shifted forward in the waveguide, and it is the length and width of the waveguide that determines the shift. There are also two Fresnel zones in the antenna, and for each Fresnel zone the signal path is one wavelength longer than the previous one. Due to the circular symmetric design, the Fresnel step can be designed with an optimized waveguide, which is repeated in this circular zone, avoiding the shadowing effect encountered in design attempts by others in past times. The Fresnel zone design demands a short waveguide outside the inner waveguide (the Fresnel step). To guide the wave around the inner longer waveguide, a dual mode home structure at entrance and exit is added to a short, narrow waveguide in between. The result is that all signals transmitted through the antenna reaches the focal point in phase. The circular symmetry of the antenna eases the control of the electromagnetic behavior and reduces cross-polarization. The media inside the waveguides is dielectric technical EPS with the effective permittivity $\varepsilon = 1{,}1$. The dielectric makes the extremely thin copper waveguide walls possible with good mechanical robustness, and it also allows a more compact construction as compared to air-filled waveguides. The hole in center was both functional, to give more open appearance and to minimize windage – in fact, an 80-cm Cybertenna behaves in wind like a 70-cm dish.

Fig. 4.4 S. Petersson and Cybertenna (Photo credit: Telesatellite magazine with permission by Alexander Wiese)

The idea of waveguide array is not novel. For example, in the paper [20] the antenna that consists of a waveguide lens illuminated by a 19-element feed-horn array was described. The antenna has a 30-in-diameter aperture and a 30-in focal length. It is made up of ≈ 700 lengths of 1 by 1 in square waveguide epoxy bonded and supported by four booms and a flange. Although the brass lens shown weighs more than 80 lbs, a titanium lightweight unit that is dimensionally identical weighs ≈ 7 lbs. The waveguide lens is a two-dimensional nonplanar array of waveguide radiators. It was shown it is necessary to step (or zone) the lens to reduce its thickness and to increase its bandwidth. A step in the lens is made whether lens thickness exceeds $\lambda/(1-\mu)$; the length of the waveguide removed increases the effective electrical path length by one or more wavelengths. The lens may be stepped on either (or both) of its surfaces. However, steps on the surface opposite the feed appear preferable in order to reduce shadowing effects. In view of the foregoing, zoning a lens has some deleterious effects on its radiation characteristics; the principal one being a reduction of efficiency, since radiation in the forward direction is greatly reduced for certain radiators. A secondary effect is to increase the H-plane beamwidth and decrease the E-plane beamwidth.

It was also mentioned since waveguides are quite dispersive, the bandwidth of a waveguide lens antenna is small. However, in a stepped lens, the phase error does not increase monotonically as a function of radial distance, but follows a stepwise increase, and the effect of such error on the secondary pattern must be obtained and the bandwidth determined from the criteria applied to the results. It is interesting to note that zoning also results in an improvement of the lens bandwidth, as short waveguides reduces the phase distortion due to dispersion.

4.6 Active Transmit Lens Array Antenna Concept

An active transmit lens array (ATLA) antenna has four major sections, three of which comprise the RF lens. The fourth is a feed array. These sections, shown in an artists sketch of a proof of concept antenna in Fig. 4.5, are:

- A flat array of radiating elements launches the far-field spot beams.
- A corresponding array of power modules, one for each radiating element, supplies signal power to the radiating array. Each power module contains an amplifier and a delay line. Collectively, the delay lines of the modules constitute the center portion of the lens.
- Signals arrive at the modules from individual receive elements at their inputs.
- The receive elements are space-fed, in turn, from an array of feed horns that is physically separate from the lens. Each far-field beam is created by the signals transmitted from a single feed horn. The beam direction is fixed by the location of the feed horn in the feed array.

The attraction of the ATLA is that no beam forming network, per se, are required. Thus, in principle, a major difficulty encountered in the active transmit phased array concept is avoided. The beam-forming network is the lens that forms all beams simultaneously, with directions determined by the geometry of the feed array and lens structure.

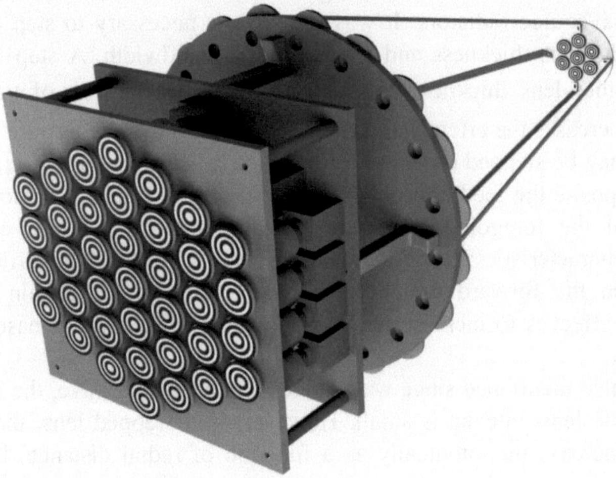

Fig. 4.5 Active transmit lens array antenna concept

4.6.1 Hexagonal Array and Active Area for Circular Elements

Hexagonal array numerology is presented followed by calculation of the active area of a hexagonal array of circular feed horns [21]. It is determined that the ratio of active area to total area is about 0.91 for large arrays.

In a hexagonal array, the total numbers of elements n for regular hexagonal fill areas are 7, 19, 37, 61, 91,127, etc. The number of elements on a principal diagonal of the hexagonal area can be defined as:

number of elements on principal diagonal = $2k + 1$,

where $k = 1$ for $n = 7$, $k = 2$ for $n = 19$, $k = 3$ for $n = 37$, etc. With this nomenclature, n may be expressed as a function of k as:

$$n = 6[1 + (k - 1)/2]k + 1.$$

Let the hexagonal fill area be inscribed within the smallest circle such that no part of the hexagonal area is outside of the circle. The diameter of this circle is equal to $(2k + 1)D$, where D is the diameter of a circular individual feed antenna.

The active area of the feed array is equal to the area of all of the n ($n = 6[1 + (k - 1)/2]\,k + 1$) feed antenna apertures, or nA, where $A = \pi(D/2)^2$ is the area of a single feed antenna aperture. The total area of the feed array is the active area plus the area taken by all of the interstitial areas between the feed antennas. Each such interstitial area A_i is obtained as follows:

A_i = (area of an equilateral triangle with vertices at the centers of three adjacent feed antennas) – 3 (one-sixth the area of a single feed antenna aperture)

$$= (1/2)D^2\cos(30^2) - (3/6)\pi(D/2)^2 = 0.161(D/2)^2 = 0.05\pi(D/2)^2.$$

It is easy to count the number of interstitial areas i for each hexagonal array and establish a functional relationship with k. The result of this effort is: $i = 6k^2$.

The ratio $R_{a/t}$ of active area to the total area of a hexagonal array is then given by:

$$R_{a/t} = \text{(active area)} /[\text{(active area)} + \text{(interstitial area)}] = nA/(nA + 6k^2 A_i) =$$

$$n\pi(D/2)^2 / \left[n\pi(D/2)^2 + 6k^2 0.05\pi(D/2)^2\right] =$$
$$(3k^2 + 3k + 1)/(3.08k^2 + 3k + 1).$$

Values for $R_{a/t}$ are given in Table 4.2.

To obtain an array with 1323 active elements, the value of k should be about 21. Therefore, the diameter of the array is 43 feed antennas. For any given feed-antenna spacing, in wavelengths, the diameter may be calculated.

For example, if antenna spacing is 2.6 wavelengths at 30 GHz, the array diameter is given by: array diameter = 43×2.6 (300/30) mm = 1.118 m.

Table 4.2 Table of values of $R_{a/t}$ for k =l, 2, …, 6

k	n (number of feeds in regular hexagonal grid array) $6[1 + (k - 1)/2]k + 1$	i (number of interstices within regular hexagonal grid array) $6k^2$	$R_{a/t}$ (ratio of active area to total area of array)*
0	1	0	1
5	91	150	0.9220
10	331	600	0.9149
15	721	1350	0.9123
20	1261	2400	0.9110
25	1951	3750	0.9102

* $R_{a/t} = (3k^2 + 3k + 1)/(3.308k^2 + 3k + 1)$.

4.7 Layered FZPL Antenna Array

As a means of improving the aperture efficiency, Jiang and Zhang [22] proposed a double FZPL (DFZPL) antenna (FZP array, which consists of two FZP). The idea was to improve the gain of the standard FZPL antenna by using the second lens to compensate for, or simply to cancel, the reflection off the metal zones from the first lens. This antenna geometry consisted of two standard FZPL antennas separated by a specific distance d as illustrated in Fig. 4.6(a).

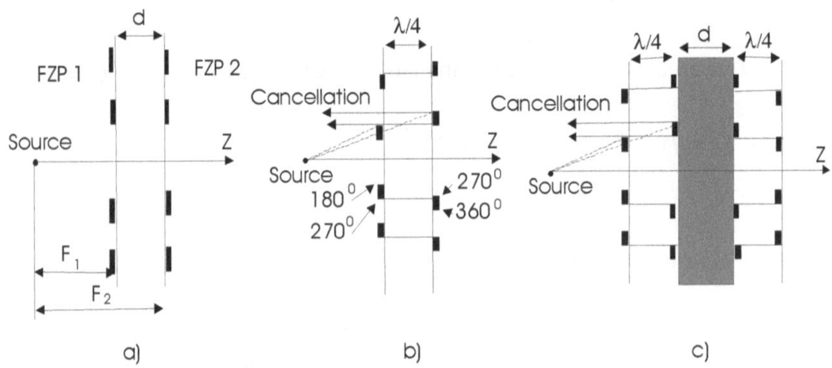

Fig. 4.6 FZP antenna arrays: (**a**) double Layer, (**b**) Enhanced Double Layer, (**c**) Four-Layer FZP Array (after [23])

Since both lenses must have the same focus, the authors defined the focal distance F_2 of the second lens to be:

$$F_2 = F_1 - d,$$

where F_1 is the focal distance of the first lens. The difference in focal lengths implies that the radii of the zones in each lens will be slightly different. The authors simplified their analysis by considering only the case where the space between the lenses was air instead of dielectric to avoid reflections from the dielectric interface. By modeling the normalized scattered field at the focus, they showed that about a 2-dB improvement in gain could be achieved by selecting odd multiple of $\lambda/4$.

Jiang and Zhang [23] also improved the backward radiation of the FZPL antenna by halving the radii of the rings on the first layer and then placing the other half on the second layer a distance $\lambda/4$ away. By doing this, the radiation will reflect as per normal off the metal rings of the first lens sub-zone with phases 180° to 270°. The radiation that reflects off the metal ring on the second layer, with phases 270°–360°, will be exactly out of phase with the reflected radiation from the first layer. The two radiations will cancel and greatly decrease the backward radiation of the antenna.

According to the ray optics, the spacing between the layers should be:

$$d = \frac{1}{L}\sum_{l=1}^{L} d_l \;; \quad \frac{1}{\varepsilon_r}\left(1+\frac{1}{\cos\theta_l}\right)d_l + \frac{1}{4}\lambda = (2n-1)\frac{\lambda}{2},$$

where ε_r is the permittivity of the media filled between the layers. The radii of the l-th ring on each FZP layer can be written as follows:

$$R'_{1l-} = \sqrt{F_2 \cdot (4l-1)\lambda/2 + ((4l-1)\lambda/4)^2}, \quad R'_{1l+} = \sqrt{F_2 \cdot (2l)\lambda + (l\lambda)^2};$$

$$R'_{2l-} = \sqrt{F_1 \cdot (2l-1)\lambda + ((2l-1)\lambda/2)^2}, \quad R'_{2l+} = \sqrt{F_1 \cdot (4l-1)\lambda/2 + ((4l-1)\lambda/4)^2}.$$

Unfortunately, the sidelobe levels were found to increase substantially in this version of the FZPL antenna array in comparison to the FZPL antenna. This structure is shown in Fig. 4.6(b).

The same authors [23] further improved the performance of the FZPL antenna array by making a four-layered version. This structure is shown in Fig. 4.6(c) and consists of two-elements FZPL antenna array with improved backward radiation separated by a specific distance.

The radii of the l-th ring on each layer are given as:

$$R_{\substack{1l-\\4}} = \sqrt{F_{\substack{1\\4}} \cdot (2l-1)\lambda + ((2l-1)\lambda/2)^2}, \quad R_{\substack{1l+\\4}} = \sqrt{F_{\substack{1\\4}} \cdot (4l-1)\lambda/2 + ((4l-1)\lambda/4)^2} \cdot$$

$$R_{\substack{2l-\\3}} = \sqrt{F_{\substack{2\\3}} \cdot (4l-1)\lambda/2 + ((4l-1)\lambda/4)^2}, \quad R_{\substack{2l+\\3}} = \sqrt{F_{\substack{2\\3}} \cdot (2l)\lambda + (l\lambda)^2} \cdot$$

Numerical results revealed a 3.1-dB gain enhancement over the standard FZPL antenna. However, this structure achieved improvement at the cost of an increase in size and complexity.

4.8 Application of a Short-Focus FZP Array to the Hartmann–Shack Sensor

Lenses are transforming the incident wave front from one to another form. For example, the front of a spherical diverging wave can be changed into a spherical converging wave front. In practice, after passing throughout the real lens the waves become aberrated. The Gaussian or paraxial optics and quasioptics are no longer valid, and such departures from the ideal Gaussian behavior are known as lens aberrations.

The Hartmann–Shack sensor is a common noninterferometric wavefront sensor, suitable for real-time applications, because it does not enclose any movable parts. The sensor consists of a lens array and a focal plane detector to capture the spot pattern from lenslets, see Fig. 4.7 for a schematic of the sensor [24]. Each lenslet focuses its portion of the wavefront to a spot in its focal plane; the distance between this spot and the optical axis of the corresponding lenslet gives the local slope of the wavefront entering the lenslet. If the wavefront has a local gradient $\Delta W(x_{i;j};\, y_{i;j})$ at the lenslet $(i;j)$, the associated spot moves laterally in the focal plane $(\Delta x_{i;j};.\ \Delta y_{i;j})$ according to the laws of geometrical optics:

$$\Delta W(x_{i;j};\, y_{i;j}) = (\Delta x_{i;j};.\ \Delta y_{i;j})/f,$$

where f is the focal length of the lenslets. The spot displacements have to be determined to reconstruct the wavefront, e.g. by a least-squares fit of the measured spot displacements to a set of orthogonal polynomials. Finally, an actuator, usually an adaptive mirror, is used in a closed-loop to cancel the wavefront aberrations out and achieve a plane wavefront. The diffraction limit of the imaging system is reached.

A Fresnel lens (FZP) could be laid out. This would have the advantage of maximum efficiency at a particular wavelength, but would be limited to a narrow band because of strong chromatic aberration [25]. Also, whenever a plane wave is incident obliquely on a zone plate, higher-order wave aberrations occur [25]. The maximum incidence angle $\psi_{i\max}$ for which the off-axis aberrations are negligible and good focusing is still possible in the paraxial approximation can be calculated by the following inequality [26]:

$$\left| \pm\sqrt{2\lambda NF}\,\frac{N\lambda}{F}\sin\psi_{\mathrm{imax}} - N\lambda\sin^2\psi_{\mathrm{imax}} \right| < \frac{\lambda}{4}.$$

Here N is the number of all open zones in the Soret zone plate.

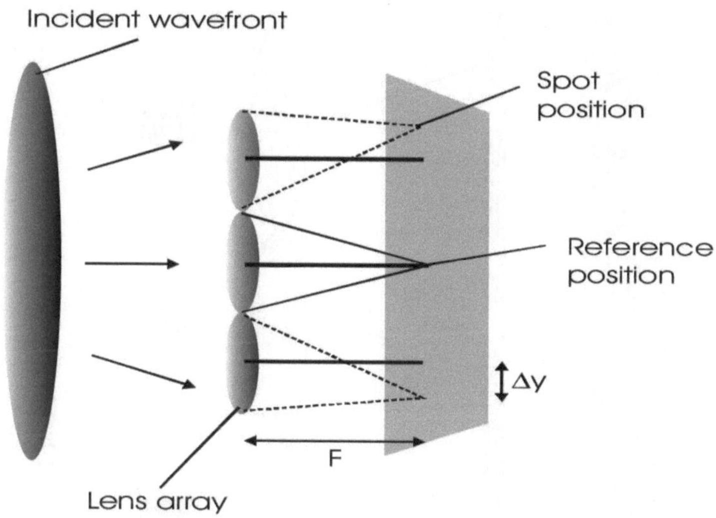

Fig. 4.7 The Hartmann-Shack sensor concept

This FZP lenslet array also must be designed to have *zero* dead space between lenses. The standard circular FZP antenna does not pack well in an array and therefore causes significant grating lobes. The HFZPL antenna avoids this problem and is therefore an attractive candidate for use in a low-profile lens array at microwave and millimeter bands. Since FZP may be a hexagonal form (see Chap. 2), there is no space in between lenses and hence the array has a 100% fill factor. For wavefront sensing, this can be important because any leakage wave will degrade the noise performance of the total system.

As it was mentioned, one disadvantage of FZP lens is a narrow band because of strong chromatic aberration. But it could be an advantage in some of cases. In combination with FZP lens array, these allow to design the concept of embedded spectrometer. The concept is shown in Fig. 4.8. Again, this FZP lenslet array also must be designed to have zero dead space between lenses.

Also these types of FZP array could be effectively applied at the multi-channel telecommunication systems of different frequency band.

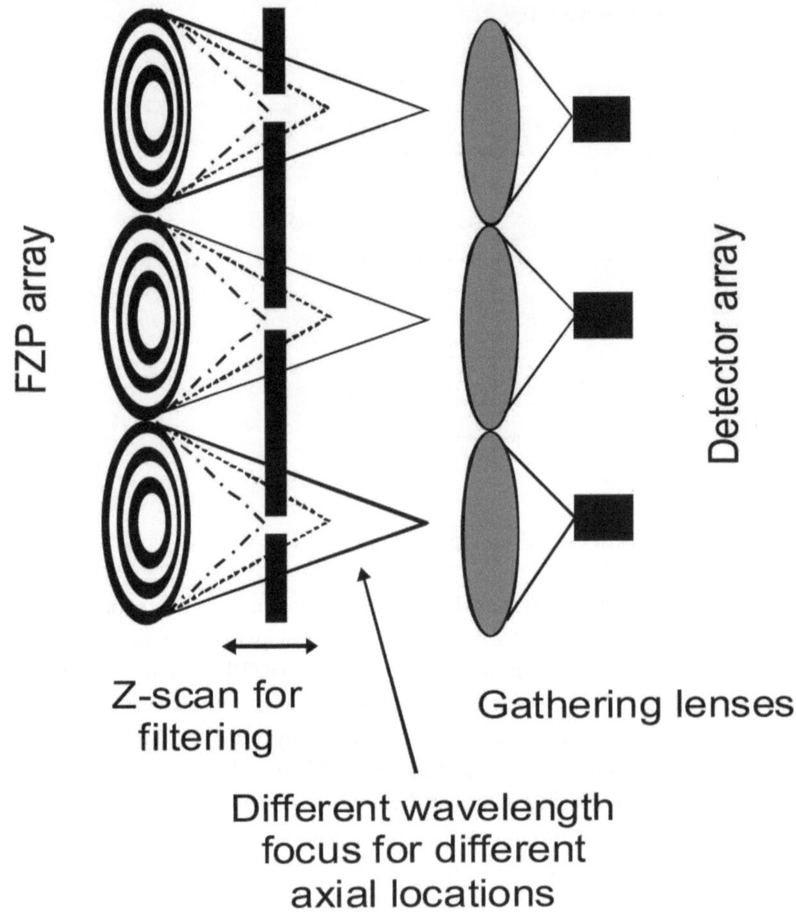

Fig. 4.8 Concept of embedded FZP array spectrometer

4.9 Exoplanet Detection Array with Unlimited Aperture

Images of high angular resolution and high dynamic range are required for the new fields of astrophysics such as exoplanet detection and cartography of stellar photospheres. Multi-aperture interferometry has been used for many years with increasing success. However, in the visible domain, interferometric arrays are still limited to small numbers of apertures.

In Sect. 2.2.4, Fresnel diffractive-interferometric array was described. Seen as a diffractive zone plate, this synthesized aperture directly forms images of high dynamic range.

The resulting wavefront quality is relatively insensitive to array warping and only limited by the precision to which the apertures are carved. This constraint is also quite loose compared to optical surfacing: a $\lambda/50$ wavefront for high dynamic range imaging requires either a $\lambda/100$ mirror or a 200 λ precision on sub-aperture edges in the Fresnel array described in Sect. 2.2.4.

But the diameter of array is limited by the precision to which the apertures rectangle elements are carved. To design a unlimited aperture, the phase function of conventional array modifies on modulo $N2\pi$ [25]. Thus, the first few Fresnel zones are conventional rectangle FZP array and classical 2π step. Then, the next several zones each incorporate two of the original Fresnel zones and have steps of 4π – see Fig. 4.9, then, the zones of several original zones and step of 6π and so on. Thus the diameter of array is unlimited.

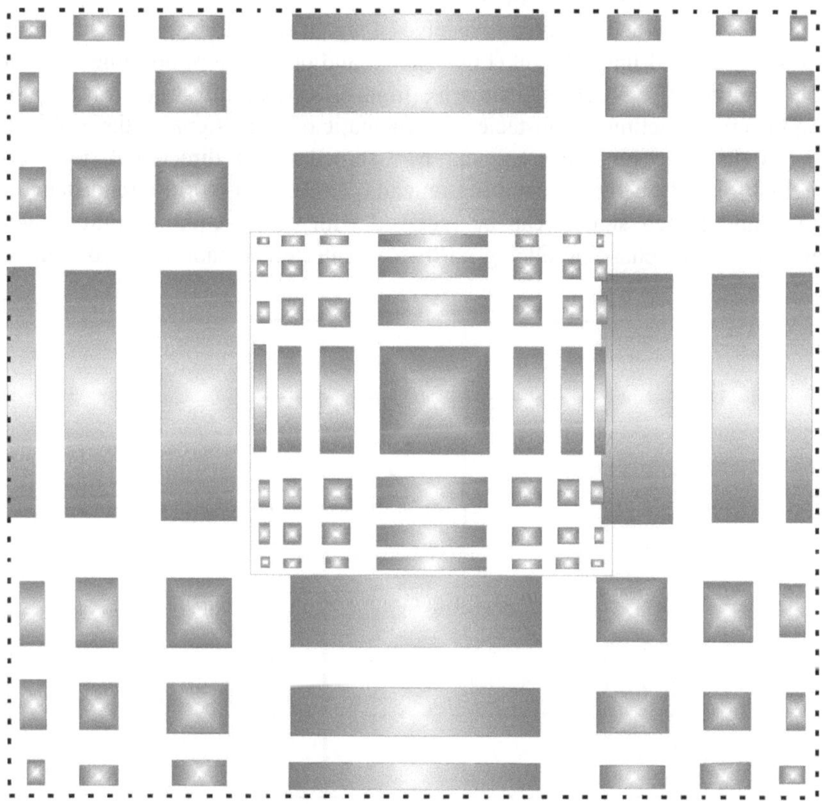

Fig. 4.9 Artist view of $N2\pi$ diffractive-interferometris array (not in scale): the solid line is a binary 2π array and dashed line is a $N2\pi$ binary array

4.10 Applications of FZP Array with Reference Phase

4.10.1 Multi-path Fading in Point-to-Point Communication

Who has not faced difficulties by a dropped mobile phone connection, dropped data packets in a wireless network, or the loss of radio reception upon stopping at a traffic light? These are all examples of the common phenomenon of multi-path fading (MPF) [27]. Multi-path effects are also responsible for ghost images in analogue television pictures and can introduce errors in a GPS receiver's locating ability. Previous attempts to eliminate the effects of MPF have not been completely successful. In the papers, however, a new technique was developed [28, 29] that can drastically reduce MPF and have shown that it can improve signal power by at least a factor of 10^4.

MPF can occur when an antenna receives a signal from a transmitter, which is a sum of the desired line-of-sight (LOS) signal and one or more non-line-of-sight (NLOS) signals. The NLOS signals arise from reflections off structures, terrain, etc. and from diffraction off obstacles. A schematic of these signal paths is shown in Fig. 4.10a. The receiving antenna is assumed to be directional, with no restrictions placed on the transmitting antenna. Under real-world circumstances, the LOS and NLOS signals can be received with nearly equal amplitude and nearly 180° out of phase producing destructive interference and a loss of carrier

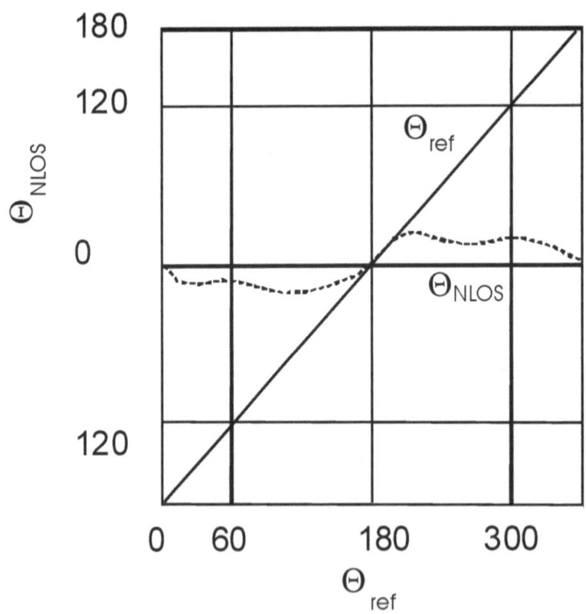

Fig. 4.10 (**a**) Schematic of two signals incident on an antenna, one is the line-of-sight (LOS) into the main beam and the other is the non-line-of-sight signal (NLOS). The LOS beam is sometimes attenuated or scattered by intervening foliage, etc. (**b**) Vector sum at feed of LOS and NLOS signals under MPF conditions

power to the receiver or MPF, as indicated in Fig. 4.10b. Historically, a variety of diversity techniques have been employed to combat MPF [30]. In these schemes, signals arriving from independent fading paths are combined in order to reduce the degree of fading, different diversity techniques including space, polarization, frequency, spread spectrum, time, and angle diversity involving two or more antennas.

Our approach to combat MPF is based on the discovery of a free parameter in the design of Fresnel zone plate (see Chap. 2) and other antennas [31]. Historically, going back to Fresnel and Soret in the 19th century, zone plates have been designed with a specific choice for this parameter that can be taken to be either a *reference radius* or equivalently a *reference phase.* It has been shown, however, that reference phase θ_{ref} can be chosen to have non-standard values, which improve the aspects of antenna performance and add new functionality to zone plate antennas. We have established both experimentally at Ka-band and theoretically that variation of θ_{ref} shifts the LOS beam phase with a positive linear slope through 360° as shown in Fig. 4.11, with only slight changes in antenna gain. Importantly, the phase of the 20° off axis NLOS signal does not change by an equivalent amount in Fig. 4.11 or can vary with a negative slope at very large NLOS angles. The difference in phase sensitivity of LOS and general NLOS signals to θ_{ref} is exploited to eliminate MPF.

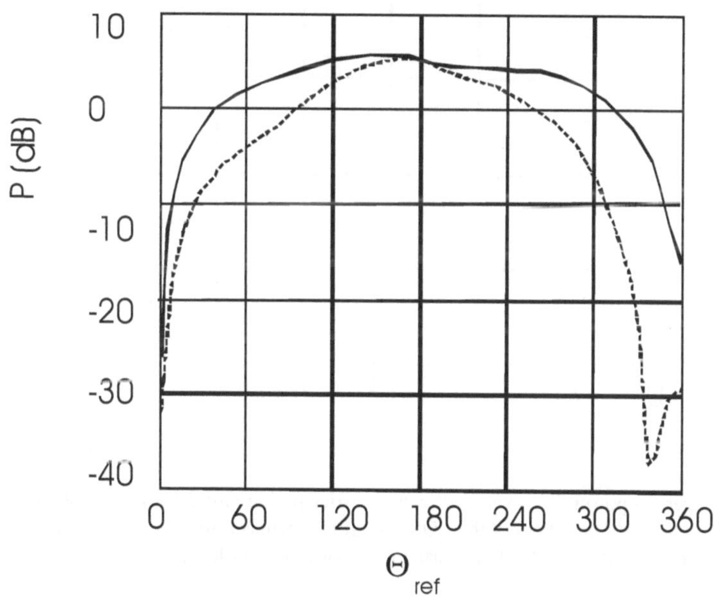

Fig. 4.11 Phases of LOS and 20° off axis NLOS signals at the feed of the FZP antenna vs. phase reference Θ_{ref}

Note that the variation of ϑ_{LOS} is 360° for a 360° variation of ϑ_{ref} both calculated and measured while the phase of the NLOS signal is much different, and that the NLOS phase variation is different from that of LOS is true for general NLOS angles.

Figure 4.12 displays the *combined* signal power of the LOS and NLOS signals of Fig. 4.11 at the receiver feed as a function of phase reference. Here, P_{avg} is the power averaged over a period of the carrier and P_{inst} is the instantaneous power at $t = 0$. In this example, the LOS signal impinging on the antenna is assumed to be attenuated by a factor of about 12 times producing a strong signal fade near $\theta_{ref} = 0°$. 0 dB is the power of the attenuated LOS signal alone. Attenuation in LOS amplitude occurs in practical situations where the LOS signal is partially obscured or scattered by foliage, terrain, etc. in outdoor settings or by walls, etc. in indoor settings. The net total power at the receiver is less than –30 dB for the standard value of θ_{ref}=0°, demonstrating strong multi-path fading. However, a static change of the phase reference increases, for example, the average power to $P_{avg} > 0$ dB for the broad range $30° < \theta_{ref} < 330°$, thus eliminating MFP.

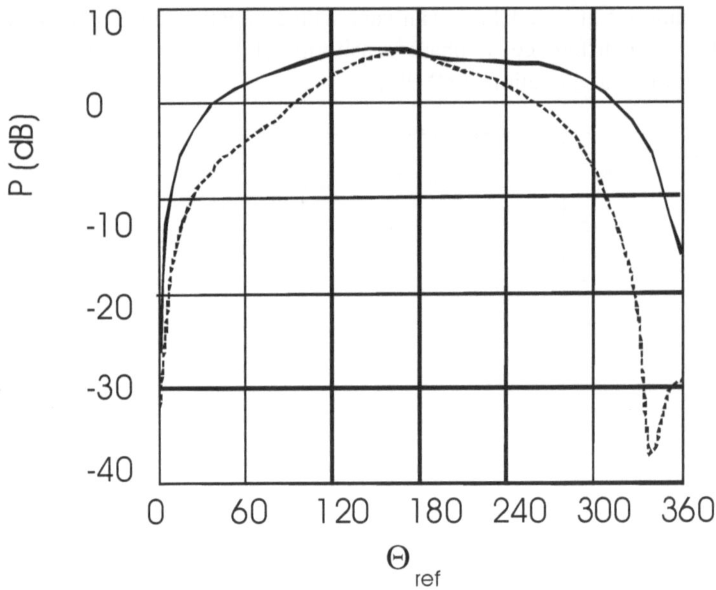

Fig. 4.12 The average power P_{avg} averaged over a carrier period (*solid line*) and instantaneous total power at $t = 0$ P_{inst} (*dotted line*) of the LOS and NLOS signals combined at the receiver feed vs. phase reference θ_{ref}. Note that the total variation of power is very large, over 30 dB

It has been demonstrated that destructive interference and thus multi-path fading can be eliminated by our new technique of using reference phase to control LOS beam phase preferentially over phase from NLOS directions [32, 33].

4.10.2 Secure Communication Using Directed Phase Modulation

The main idea is to send information by using reference phase to modulate signal phase in the main beam. The same phase modulation does not appear in off-axis directions. This could prevent eavesdropping from off-axis directions. The calculations below were made by Dr. G. W. Webb.

Below we treat a Fresnel zone plate transmit antenna that is designed to have its main lobe a beam on-axis. The reference phase of the antenna is varied from 0° to 360° and the power and phase of the signal in the main lobe and side lobe directions are calculated. It was shown that the power in the main lobe is nearly constant as a function of reference phase while the signal phase accurately tracks the reference phase through 360° [29, 33]. Thus, control over reference phase allows for signal phase modulation and the transmission of information on a constant power beam.

In off-axis directions, conversely, the signal power is reduced by 20–40 dB, and signal power varies by 5–20 dB as antenna reference phase varies through 360° – see Fig. 4.13a. Note that for the main lobe 0° direction there is one-to-one correspondence between reference phase and signal phase. Therefore, reference phase phase-modulates the phase of the beam. Also, note that for off-beam directions 10°–40° the signal phase does not follow the reference phase. Therefore, the phase modulation of the main beam is not transmitted in the off-axis directions. Importantly, the variation of signal phase in sidelobe directions is much less than in the main lobe or varies nonmonotonically and in the opposite sense from the main lobe. Thus signal information in the main lobe, encoded by reference phase, is not transmitted into the sidelobes where it could be intercepted by an eavesdropper.

Also importantly, note that signal power for the 0° direction is nearly constant near 0 dB as a function of reference phase (Fig. 4.13b). Also, note that for off-beam directions 10°, 20°, 30°, 40°, the signal power is approximately –20 to –30 dB that of the main lobe and much more strongly varying than for the main beam.

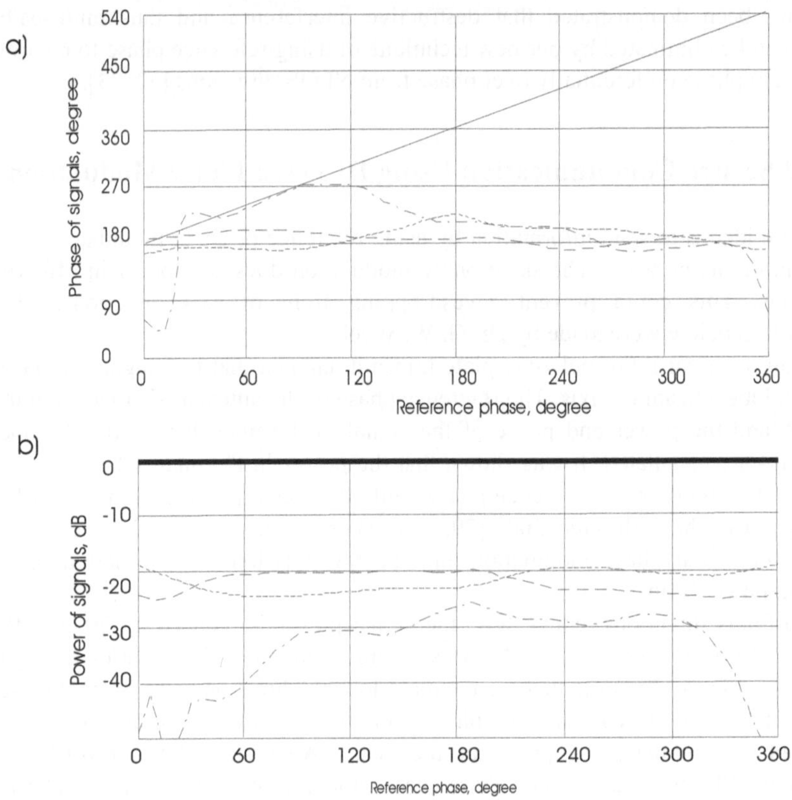

Fig. 4.13 Phase of the signal (**a**) and power of the signal (**b**) for directions of 0°, 10°, 30°, 50° as reference phase is varied from 0° to 360°

These effects may be applied to any different field of applications [34, .35].

4.11 Suggestions for Future Work

We conclude with a brief discussion on the future of Fresnel zone array antennas.

As is well known, *a phased array antenna* is composed of lots of radiating elements, each with a phase shifter. Beams are formed by shifting the phase of the signal emitted from each radiating element to provide interference so as to steer the beams in the desired direction. The main advantages with phased array antennas are advanced beam-forming capabilities and extreme beam agility when scanning. But phased array antennas are a very complex structure, making them expensive. There are also limitations in what frequency it can be used, since the antenna elements are to be located half a wavelength from each other to avoid grating lobes. For example, for 10 GHz the distance is approximately 12.5 mm and for 150 GHz it is 0.83 mm.

Phased FZP array lenses (Fig. 4.14) comprising the diffractive lens technology is an innovative approach to address the major drawbacks of a traditional phased array antenna, i.e. complexity and upper frequency limitations. By replacing the antenna elements with diffractive lenses, the number of phase shifters and other components can be reduced by a factor 1:500. Placed behind the diffractive lenses are mechanically or electronically maneuverable feed units. When placed behind the diffractive lens, the shadowing effects from the feed units on the incoming wave is eliminated. A phased array lens antenna is suitable in EHF environments as well as in higher or lower frequencies.

Moreover, if we fixed a beam position, we do not need to use phase-shifted devices because we may have to control a phase shift by means of reference phase. So the diffractive lens antenna is also a phase shifter and frequency filter itself (Fig.4.15).

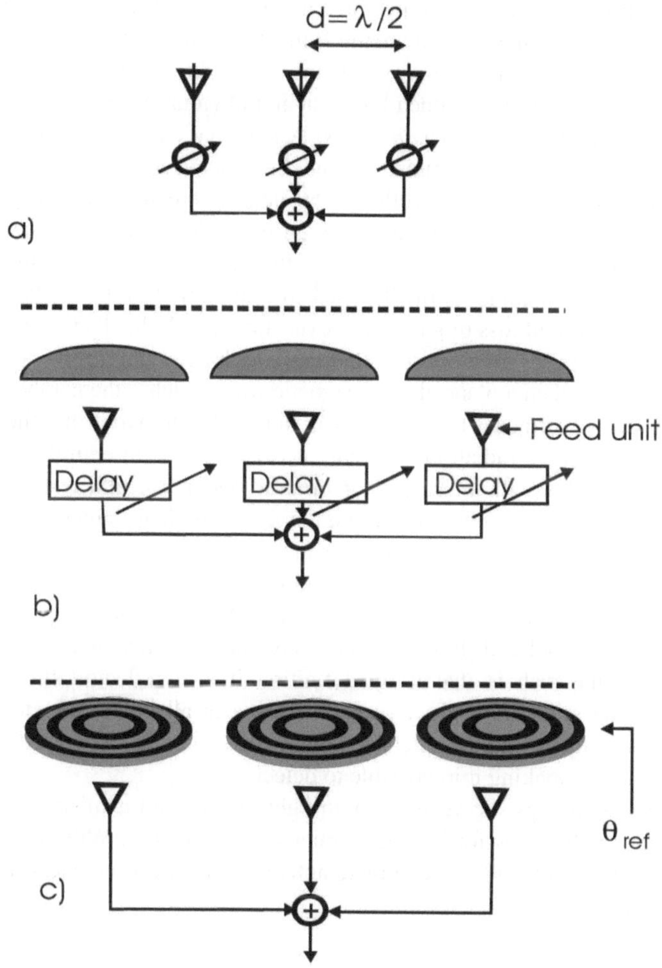

Fig. 4.14 Phased array antenna principles

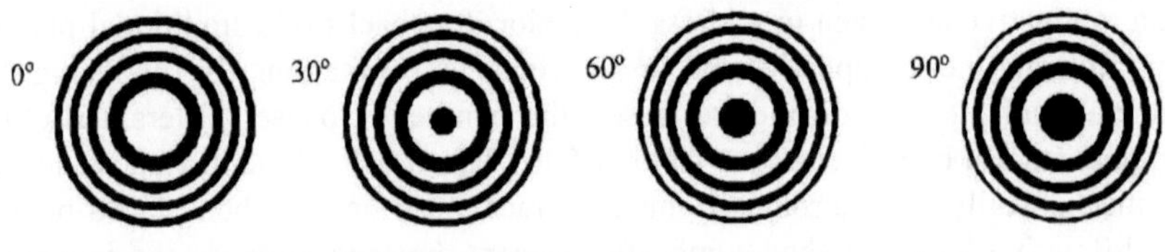

Fig. 4.15 Examples of binary masks of the FZPs with different reference phase

The diffractive lens antenna is based on a modulation of the phase, and hereby the bandwidth is limited. Signals outside the chosen frequency range are automatically filtered. This in-built electromagnetic filter improves the signal-to-noise ratio and improves bit error rate.

Also diffractive lens antenna saves a possibility of operation up to the focal length of half of wavelength, so that such an antenna is "super flat". Taking into account that lens antenna as a rule made from dielectric material, it is a high-gain stealth microwave/millimeter wave antenna. In addition, the FZP array beam width should be very narrow, much less than an individual FZP.

High gain stealth microwave antennas: A key challenge in stealth technology is to keep the wireless communication link between platforms and units undetected. Common dishes "a reflector by nature" are frequently used in such communication links due to the need of high gain. The major disadvantage with dishes in a stealth communication link is the fact that structural materials or specially designed coatings normally used to reduce radar reflectivity cannot be used on dishes without loss of gain. Moreover, for a parabolic dish, there is always the problem with noise from the surroundings which enter the receiver or feed part located in focus. When a plane electromagnetic wave reaches the parabolic surface of the dish and is refracted to a focus, it is impossible to avoid that the reflected rays cross the incident radiation. Thereby, it is impossible to entirely encapsulate the feed unit together with the antenna in order to obtain a stealth system.

Stealth antenna systems. With the diffractive lens antenna, a stealth communication system can be made, preventing the units to be detected. A key element in the stealth environment is to achieve stealth communication links in order not to reveal, for instance, a units position. Due to an increasing demand for bandwidth, a parabolic dish is the most common antenna used. But the big drawback for the dish is that it by definition it is a reflector, thereby easily detected on foe radar. A diffractive lens antenna is all but a reflector, since it focuses the signals behind the antenna. This enables to completely encapsulate the complete antenna, making it impossible to detect.

The lens technology, however, is a straightforward and available technology, solving the stealth issues for high-gain microwave antennas. With the feed units placed *behind* the antenna, the complete antenna system can be encapsulated and thereby completely stealth.

The lens technology, however, is a straightforward and available technology, solving the stealth issues for high-gain microwave antennas. With the feed units placed *behind* the antenna, the complete antenna system can be encapsulated and thereby completely stealth.

4.11.1 Zones Structure with Overlapping

One of the big problems with arrays of classical high-gain elements is that they have a very limited beam scan range capability, due to the appearance of grating lobes because of the wide element spacing.

For example, Fig. 4.16 shows simulated element pattern for two types of 2×2 FZP arrays. Two-zone case shows that part of the array factor first sidelobes are outside of the element pattern. This is even more so for the three-zone case.

We know that the conventional axisymmetric zone plate with circular zones forms a beampattern perpendicular to the reflecting surface. If we now "juxtapose" the amplitude masks of two zone plates, new areas of jointly formed or conformal zones will appear in the area of intersection of transparent and opaque zones. Fig. 4.17 shows such a "composite" zone plate that joins two zone plates with overlapping. Obviously, the new zone plate obtained by this operation generates simultaneously two different beam patterns [25, 36, 37].

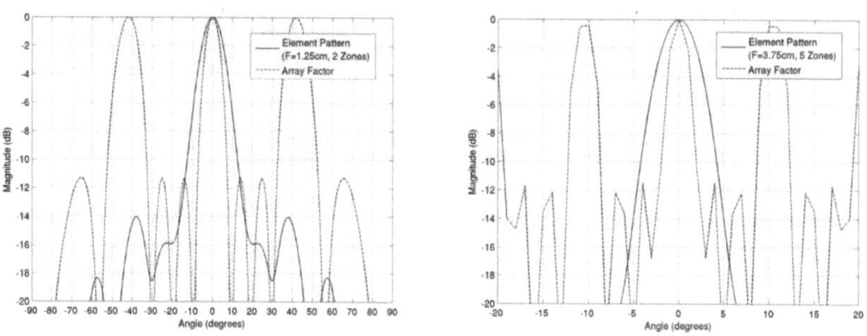

Fig. 4.16 Element pattern (*solid line*) and array factor of 2×2 array (*dashed line*). Left – HFZP with $F = 1.25\,\lambda$, 2 metal zones with 3λ element spacing on the equilateral triangle, right – HFZP with $F = 3.75\,\lambda$, 5 metal zones with $11\,\lambda$ element spacing on the equilateral triangle (with permission by Dr. Sara M. Stout-Grandy)

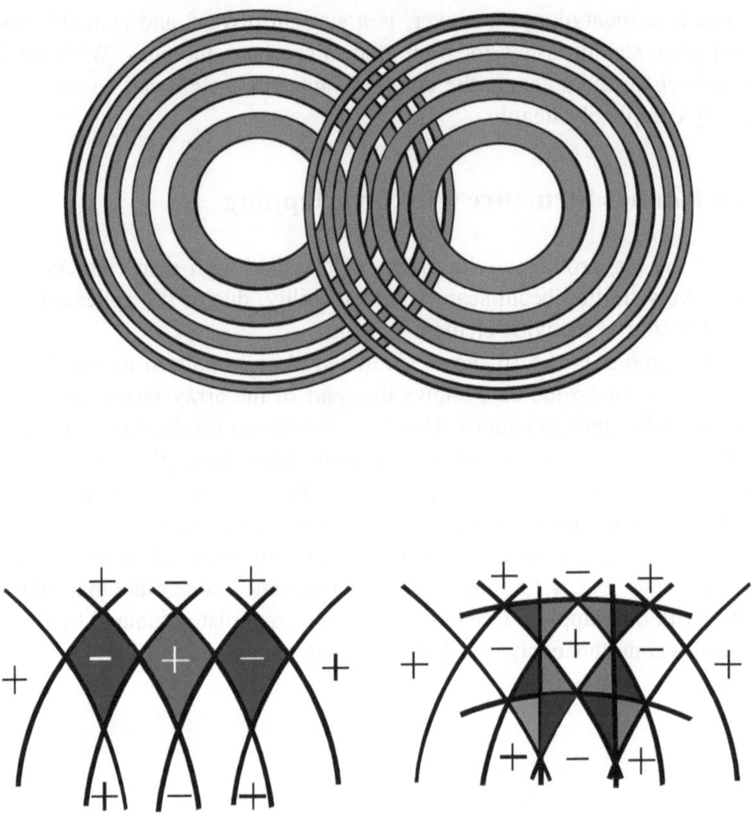

Fig. 4.17 The structure of conformal zones (after [37] with kind permission of Dr. J. Vassal'lo)

Mathematically, the zones intersection may be found following to algorithm [25, 36, 37]. It is easy to show from the geometrical consideration that the points of intersection of circular boundaries are located according to the following equations:

$$x_n = (d^2 + r_n^2 - R_m^2)/2d\,,\ y_n = \sqrt{r_n^2 - d^2/4}\,,$$

where d is the distance between centers of FZPs, $0 \le d \le R + r$. In the case of identical FZP, i.e. $r_n = R_n$ (or FZPs with equal reference phase), the correspondent equations are:

$$x_n = d/2,\ y_n = \sqrt{r_n^2 - d^2/4}\,.$$

In this method, the united phase function of conformal FZP could be determined.

4.11.2 The Main Areas for Further Research

In the conclusion, it appears that there are at least five main areas for further research.

First, the application of the Fresnel zone concept and novel phase-correcting techniques to various types of antennas, such as sub-wavelength focus antenna, including photon crystal lens, described briefly in Chap. 3, and photonic crystal substrate and reflector.

Second, the development of efficient and low-cost phase-correcting techniques with reference phase control optimization.

The third area is the development and study of FZP array with conformal Fresnel zones.

The fourth area is the study of conformal FZP array on arbitrary 3D surface.

The fifth area is the study of broadband multi-beam Fresnel zone array antennas. At the moment, the telecommunications industry is moving toward the UMT Service and other broadband systems, and there is a high demand for novel antenna solutions to satellite, cellular mobile, and WLAN radio problems. Moreover, the automotive industry is introducing more and more radio technology to make driving a safer and more enjoyable experience. And low-cost and broadband Fresnel zone array antennas including dynamically variable zones structures will find many applications in those areas, including terrestrial and satellite communication systems and millimeter-wave ground-based communications. Cellular telephone and data link infrastructure are by far the largest commercial applications of millimeter-wave systems.

References

1. E. Brookner (ed.). Practical Phased Array Antenna Systems. Artech House, Boston MA, 1991.
2. R. J. Mailloux. Phased Array Antenna Handbook. Artech House, 1994.
3. J. C. R. Poirier, G. A. Morin, Y. M. M. Antar, J. W. Moffat, Millimetre-Wave Limited-Scan Array Using Small Lenses. Antennas and Propagation Society International Symposium, 2001. IEEE Vol. 2, 2001 pp.:823–826.
4. L. Shafai, D. J. Roscoe, M. Barakat. *Simulation and Experimental Study of Microstrip Fed Cavity Antennas*, ANTEM, pp. 549–554, August 1996.
5. W. B. Dou, G. Zeng, Z. L. Sun, Pattern Prediction of Extended Hemispherical-Lens/objective-Lens Antenna System at Millimetre Wavelengths. IEE Proc.-Microw. Antennas Proa., Vol. 145, No. 4, .pp. 295–298, August 1998.
6. G. V. Eleftheriades, Y. Brand, J. -F. Ziircher, J. R. Mosia, *ALPSS:* A Millimetre- Wave Aperture-Coupled Patch Antenna on a Substrate Lens. Electronics Letters, Vol. 33, No. 3, January 1997 pp. 169–170.
7. J. R. James, C. John, C. M. Hall. Millimetre-Wave Hybrid Dielectric-Microstrip Antenna Array. IEE Proc. H., Microwaves, Opt. & Antennas, Vol. 131, 1984, pp. 341–350.
8. W. F. Crosswell, J. S. Chatterjee, V. B. Mason, C. T. Tai. Aperture Excited Dielectric Antennas' NASA Tech. Note, 1974, TN D-7342.

9. C. M. Hall, G. Anderasic, J. R. James. Microstrip Planar Arrays with Dielectric Sphere Overlays. Electronics Letters, Vol. 21, No. 8, April 1985, pp. 356–357.
10. R. Chatterjee. Dielectric and Dielectric-Loaded Antennas. Wiley, 1985, Chapter 7.10 "Microstrip Patch Arrays With Spherical Dielectric Overlays".
11. R. Beresford, A. Chippendale, D. Ferris, P. Hall, C.e Jackson, G. James, M. Wieringa. Eyes on the Sky: A Refracting Concentrator Approach to the SKA. The Executive Secretary Australian SKA Consortium Committee PO Box 76, Epping, NSW 1710, Australia. June 2002. 109p.
12. L. J. Ricardi, F. W. Cipolla. Low Profile Hemispherical Lens Antenna Array on A Ground Plane. US Patent 005781163A, Jut 14, 1998.
13. D. J. Raponi, K. Loucks, P. J. Oleski, D. J. Cobb. Airborne Demonstration of Milstar and GBS Receive Capability Using a Single Antenna. The MITRE Corporation, www.uhf-satcom.com/kaband/airborneGBS.pdf
14. J. Jahns, S. J. Walker, .Two-Dimensional Array of Diffractive Microlenses Fabricated by Thin Film Deposition. Applied Optics, Vol. 29, No. 7, March 1990, pp. 931–936.
15. T. H. Bett, C. N. Dnason, P. Jinks, D. A. Pepler, I. N. Ross, R. M. Stevenson. Binary Phase Zone-Plate Arrays for Laser-Beam Spatial-Intensity Distribution Conversion. Applied Optics, Vol. 34, No. 20, July 10, 1995, pp. 4025-4036.
16. R. B. Dybdal. Millimeter Wave Antenna Technology. IEEE Journal on Selected Areas In Communications, Vol. SAC-1, No. 4, September 1983, pp. 633–644.
17. N. Ya. Molotkov. Study of focusing and scattering systems with variable refractive index in the university course of optics. Deposition no 114-B87, Tomsk, 1987, 11 pp. Board of "Izvestiya vuzov MV i SSO SSSR", seriya "Fizika".
18. S. Petersson. Development of a Mass-Producible Waveguide lens antenna. Proc. of the 34th European Microwave Conference – Amsterdam. 2004, pp. 1197–1199.
19. S. Petersson. US Patent 6.426.728 B2, Jul.30, 2002. Patent No 510 565 "Waveguide lens".
20. A. R. Dion, L. J. Ricardi. A Variable-Coverage Satellite Antenna System. Proceedings of the IEEE, Vol. 59, No. 2, February 1971, pp. 252–262.
21. W. C. Mitchell, L. Nguyen, A. Dissanayake, B. Markey, A. Le. Rain Fade Compensation for Ka-Band Communications Satellites. NASA/CR-97-206591
22. G. Z. Jiang, W. X. Zhang, The Effect of Layer Spacing on the Properties of Double-Layer Fresnel Zone Plate Lens. IEEE Antennas and Propagation Symposium, 1997, pp. 472–475.
23. G. Z. Jiang, W. X. Zhang, T. H. Liu. New Types of Fresnel Zone Plate Lens with Lower Backward Radiation. IEEE Antennas and Propagation Symposium, 1998, pp. 166–169.
24. L. N. Thibos. Principles of Hartmann-Shack Aberrometry. In Trends in Optics and Photonics Vol. 35, Vision Science and Its Applications, V. Lakshminarayanan, Ed., Optical Society of America, Washington, DC, pp. 163–169, 2000.
25. O. V. Minin, I. V. Minin. Diffractive Optics of Millimeter Waves. – IOP Publisher, Boston-London, 2004. 396p.
26. M. S. Raiski, Zone Plate, Uspehi Phys. Nauk, Vol. XLVII, 4, August 1952, pp. 515–536 (in Russian).
27. http://wireless.per.nl/reference/chaptr03/fading/fading.htm
28. G. W. Webb, I. V. Minin, O. V. Minin. Elimination of multipath fading in point-to-point communication// Proceeding of the 5th IEEE-Russia Conference "Microwave electronics: measurements, identification, applications" MEMIA 2005, December 13–15, 2005, Novosibirsk, Russia, pp.142–146.
29. G. W. Webb, I. V. Minin, O. V. Minin. New technique to combat multipath fading in wireless networks. Proc. SPIE 6248 Wireless Sensing and Processing, Raghuveer M. Rao, Sohail A. Dianat, Michael D. Zoltowski, Editors, 62480P (May. 12, 2006), 17–21 April 2006 Gaylord Palms Resort and Convention Center, Orlando (Kissimmee), Florida USA, pp. 24–27.
30. G. L. Stuber, Principles of Mobile Communication, 2nd Ed., Kluwer, Norwell. 2001.
31. G. W. Webb, I. V. Minin, O. V. Minin, in preparation.

32. G. W. Webb, I. V. Minin, O. V. Minin. Eliminating multipath fading improves wireless signal reception. SPIE newsroom. Communications and Networking, 2006, ID 10.1117/2.1200607.0269
33. G. W. Webb, I. V. Minin, O. V. Minin. New Technique to Suppress Sidelobe Clutter in Perimeter Security Systems. International Journal of High Speed Electronics and Systems (IJHSES) – Volume 17, No. 2, June 2007, pp.. 367–382.
34. G. W. Webb, I. V. Minin, O. V. Minin. Technologies of Millimeter-Wave Road – Vehicle and Vehicle – Vehicle Communications. Digest of the Joint 31st International Conference on Infrared and Millimeter Waves and 14th International Conference on Terahertz Electronics, September 18–22, 2006, Shanghai, China, p. 357.
35. I. V. Minin, O. V. Minin. Array of Fresnel Zone Plate Lens Antennas: Circular, Hexagonal with Chiral Symmetry and Hexagonal Boundary. Digest of the Joint 31st International Conference on Infrared and Millimeter Waves and 14th International Conference on Terahertz Electronics, September 18–22, 2006, Shanghai, China, p. 270.
36. A. Pedreira, J. Vassal'lo. Patent solicitude no 200001599 (E), Spain.
37. I. V. Minin, O. V. Minin, A. Pedreira, J. Vassal'lo. Some Fundamental Principles of the Fzp-Like Antenna Developments. Int. Conf. On Actual Problems on Electronics Instrument Engineering, Proceeding, APIEE-2002, Vol. 1, pp.176-177, Novosibirks, September 23–26.

MIX
Papier aus verantwortungsvollen Quellen
Paper from responsible sources
FSC® C105338

If you have any concerns about our products,
you can contact us on
ProductSafety@springernature.com

In case Publisher is established outside the EU,
the EU authorized representative is:
Springer Nature Customer Service Center GmbH
Europaplatz 3, 69115 Heidelberg, Germany

Printed by Libri Plureos GmbH
in Hamburg, Germany